Metaheuristics: Outlines, MATLAB Codes and Examples

Ali Kaveh • Taha Bakhshpoori

Metaheuristics: Outlines, MATLAB Codes and Examples

Springer

Ali Kaveh
School of Civil Engineering
Iran University of Science
and Technology
Narmak, Tehran, Iran

Taha Bakhshpoori
Faculty of Technology and Engineering
Department of Civil Engineering
University of Guilan, East of Guilan
Rudsar, Vajargah, Iran

ISBN 978-3-030-04066-6 ISBN 978-3-030-04067-3 (eBook)
https://doi.org/10.1007/978-3-030-04067-3

Library of Congress Control Number: 2018962367

This Springer imprint is published by the registered company Springer Nature Switzerland AG.
The registered company address is: Gewerbestrasse 11, 6330 Cham, Switzerland

Preface

Optimization problems arise in various disciplines like engineering, economics, and sciences. In view of the practical utility of these problems, there is a need for efficient and robust solvers. Metaheuristic algorithms as probabilistic solvers are developed and considered as the primary choice instead of deterministic approaches. A brief review of the metaheuristic-based optimization literature is clear proof to this fact: metaheuristics are evolving faster and faster. The dual nature (random based and rule based) of these algorithms makes them efficient to tackle hard optimization problems in a reasonable time with acceptable accuracy, however, this property makes their characterization very difficult.

Many books have been published in this area on theory (from theoretical properties to performance guarantees), computational developments (from empirical analyses to extending the available varieties), and applications (tackling ever more challenging problems in various disciplines). Many of them are overviews of the most widely mature algorithms such as evolutionary algorithms, tabu search, ant colony algorithm, particle swarm optimization, and harmony search algorithm. In recent years, novel efficient and increasingly employed algorithms with different roots have been developed one after another. There appears to be a need for presenting a selective set of recent metaheuristic algorithms with practical treatment to a broader audience, ranging from undergraduate science and engineering students to experienced researchers and scientists as well as industry professionals. Such a framework can be useful for the application of metaheuristics in hard optimization problems, comparing conceptually different metaheuristics and designing new ones.

This book is an attempt to present five recently well-known and increasingly employed algorithms besides eight novel and efficient developed algorithms by the first author and his graduate students in an accompanying practical implementation framework.

The framework is supplied with MATLAB codes and one benchmark structural optimization problem. The aim is to provide an efficient context for experienced researchers or those readers not familiar with theory, applications, and computational developments of the considered metaheuristics. This framework can easily be further extended to existing improved or hybrid versions of the algorithms and solving new problems.

The readers interested in the application of metaheuristics for hard optimization, comparing conceptually different metaheuristics and designing new metaheuristics, are invited to read this book. The book is likely to be of interest to civil, mechanical, industrial, and electrical engineers who use optimization methods for design, as well as to those students and researchers in structural optimization who will find it to be necessary professional reading.

In Chap. 1, a short introduction is provided for the goals and contents of this book. Chapter 2 provides the required preliminaries and presents the frameworks needed to understand the subsequent chapters. Chapter 3 provides optimum design by artificial bee colony algorithm. In Chap. 4, Big Bang-Big Crunch algorithm is presented. Chapter 5 contains the teaching-learning-based optimization algorithm. Chapter 6 presents imperialist competitive algorithm which is a socio-politically motivated optimization algorithm. Chapter 7 presents an efficient population-based metaheuristic inspired by the behavior of some cuckoo species in combination with the Lévy flight. Chapter 8 contains charged system search algorithm which utilizes the governing Coulomb laws from electrostatics and Newtonian laws of mechanics. In Chap. 9, the ray optimization algorithm is presented which is conceptualized based on the Snell's light refraction law when light travels from a lighter medium to a darker medium. In Chap. 10, colliding bodies optimization algorithm is studied which is inspired by the physics laws of momentum and energy governing the collisions occurring between solid bodies. Chapter 11 deals with the presentation of tug of war optimization algorithm inspired by the game of tug of war where each candidate solution is considered as a team participating in a series of rope pulling competitions. In Chap. 12, water evaporation optimization is discussed which is inspired by evaporation of a tiny amount of water molecules on a solid surface with different wettability that can be studied by molecular dynamics simulations. In Chap. 13, vibrating particles system algorithm is presented which is motivated based on the free vibration of single degree of freedom systems with viscous damping. Chapter 14 presents cyclical parthenogenesis algorithm that is inspired by reproduction and social behavior of some zoological species like aphids, which can reproduce with and without mating. In Chap. 15, thermal exchange optimization algorithm is discussed which is based on the Newton's law of cooling.

We would like to take this opportunity to acknowledge a deep sense of gratitude to a number of colleagues and friends who in different ways have helped in the preparation of this book. Our special thanks are due to Ms. Silvia Schilgerius, the senior editor of the Applied Sciences of Springer, for her constructive comments, editing, and unfailing kindness in the course of the preparation of this book. Our sincere appreciation is extended to our Springer colleagues, Mrs. Rajeswari Rajkumar (project manager) and Mrs. Sujatha Chakkala (production editor), who prepared the careful page design of this book. We would also like to thank Dr. M. Ilchi Ghazaan and A. Dadras for their valuable suggestions and M.R. Seddighian and A. Eskandari for proofreading some chapters of this book.

Every effort has been made to render the book error free. However, the authors would appreciate any remaining errors being brought to their attention through the following email addresses:

alikaveh@iust.ac.ir (Prof. A. Kaveh) and tbakhshpoori@guilan.ac.ir (Dr. T. Bakhshpoori)

Tehran, Iran
October 2018

Ali Kaveh
Taha Bakhshpoori

Contents

Chapter 1
Introduction

1.1 Optimization

Optimization is a part of nature and, inevitably, an integral part of human life. Any decision we make is an attempt to process an optimal or almost optimal situation. From a general point of view, any optimization problem can be considered as a decision-making problem, and the question is whether or not there is any better solution to the problem than the one we have found. In other words, optimization means to reach a solution as good as possible to lead us to a better performance of the system under consideration. According to the description of Beightler et al. [1], optimization is a three-step decision-making process: (1) modeling the problem based on the knowledge of the problem, (2) finding measures of effectiveness or the objective function, and (3) the optimization method or optimization theory. It is valid to say that the entire field of optimization, especially the latter step, is merely benefited by the development and improvement of computers which started in the mid-1940s.

The existence of optimization methods can be traced to the days of Newton, Bernoulli, Lagrange, Cauchy, and Gibbs in which mathematical analysis was created on the basis of the calculus of variations [2]. Such optimization methods are generally known as mathematical programming and encompass a great deal of advanced literature over the decades. With the advent of computers, the need for optimizing more complicated and inherently nonlinear optimization problems came into the scene and resulted in new advances in optimization theory. Another area of optimization theory known as metaheuristics was developed in the mid-1940s and received considerable attention in the last half a decade which is the main focus of this book.

As opposed to mathematical programming techniques which guarantee optimality and metaheuristics, which are known also as approximate or nontraditional methods, aim to find optimally acceptable solutions for the optimization problem in a practically reasonable time. It should be noted that metaheuristics should not be

A. Kaveh, T. Bakhshpoori, *Metaheuristics: Outlines, MATLAB Codes and Examples*, https://doi.org/10.1007/978-3-030-04067-3_1

confused with approximation algorithms. In technical sense, a metaheuristic tries to combine basic heuristic methods in higher-level frameworks with the aim of exploring effectively a search space. In other words, metaheuristics try to merge randomization and rule-based theories which are almost always taken from natural phenomena such as evolution, characteristics of biological systems, social systems, swarm intelligence, and governing laws in different phenomena like basic physical laws. The term metaheuristic was introduced for the first time by Glover [3] and has been extensively used for this type of algorithms. The word metaheuristic is a combination of two old Greek words: the verb *heuriskein* and the suffix *meta* which mean "to find" and "beyond, in an upper level," respectively. A unified definition is not available in the literature. The definition by Osman and Laport [4] is addressed here: "A metaheuristic is formally defined as an iterative generation process which guides a subordinate heuristic by combining intelligently different concepts so as to explore and exploit the search space; learning strategies are employed to structure information in order to find efficiently near-optimal solutions."

1.2 Brief Description of the Components of the Metaheuristics

There are some different criteria to classify metaheuristics, the most common of which is the population-based search versus single-solution-based search. Defining these two classes can be useful to become familiar with the metaheuristics. Single-solution-based metaheuristics manipulate and transform a single solution to obtain an optimal one applying an iterative search procedure (generation and replacement). Generation means the production of one candidate solution or a set of candidate solutions from the current solution based on higher-level frameworks or mechanisms. Replacement means the newly generated solution or a suitable solution from the generated set is chosen to replace the current solution with the aim of entering the promising region of the search space. This iterative process continues until a stopping criterion is satisfied. In the population-based metaheuristics, an iterative search procedure (containing generation and replacement) is also applied, but in this type of metaheuristics, a set of solutions spread over the search space. Firstly, a set of solutions known as initial population are initialized. Initializing can be made using different strategies, but the most common one is random generation of agents over the search space. The algorithm manipulates the current set of solutions repeatedly based on higher-level search frameworks or mechanisms to generate new ones and replace the old ones with newly generated ones using a specific strategy. This process continues until the stopping criterion is fulfilled. A fixed number of algorithm iterations, a maximum number of iterations without progress in the objective function, and a minimum value of the objective function are the most common termination criteria.

Designing a new framework as a new metaheuristic is in fact introducing new formulations for generation and replacement stages. Almost all of these formulations are inspired from natural phenomena as outlined in the previous section and characterize the performance of metaheuristics in two aspects: the ability to probe the neighborhood of previously visited promising regions and the capability to visit and probe entirely new regions of the search space. These two conflicting abilities are called exploitation (intensification or local search) and exploration (diversification or global search), respectively. An algorithm with a good balance between these two will lead to a good metaheuristic. Concerning the existing knowledge on theoretical, application, taxonomy, and survey in metaheuristics, it is rarely discussed how a balance between exploration and exploitation can be achieved. Almost in all of the developed algorithms until now, achieving a balance between exploration and exploitation has been managed by algorithmic parameter tuning, where the search for the best setting of the control parameters is viewed as an optimization problem. In this regard, recently some self-adaptive metaheuristics are invented. It should be mentioned that a large number of research works are also carried out to improve these aspects of exciting metaheuristics using different strategies such as hybridization of two algorithms or adding a mechanism to an algorithm.

One can summarize that a metaheuristic is a higher-level framework, as far as it is not problem-specific or adaptable to specific problems with few modifications, can efficiently guide the search process in complicated nonlinear and nonconvex search spaces toward near optimal solutions. Efficient guidance takes place using rules, formulations, or mechanisms to make an effective balance between exploration and exploitation.

1.3 History of Metaheuristics

It is widely accepted that metaheuristics have gained the most widespread success and development among other methods in solving many practical optimization problems. Regardless of the rich literature on the modifications of current metaheuristics using different mechanisms and strategies which have been under way, this field is witnessing the advent of new metaheuristics, perhaps once in a month. Sorensen et al. [5] very recently described the history of metaheuristics in five distinct periods: "(1) pre-theoretical period (until c. 1940), during which heuristics and even metaheuristics were used but not formally introduced, (2) early period (c. 1940–c. 1980), during which the first formal studies on heuristics appear, (3) method-centric period (c. 1980–c. 2000), during which the field of metaheuristics truly takes off and many different methods are proposed, (4) framework-centric period (c. 2000–now), during which the insight grows that metaheuristics are more usefully described as frameworks, and not as methods, (5) scientific period (the future), during which the design of metaheuristics becomes a science rather than an art." They believe that annotated and chronological list of metaheuristics as a straightforward routine of expressing history can be useful but makes no insight into the development of the field as a whole. This section does not present such a list but aims to make necessity-based distinctions to the history of metaheuristics.

Prior to 2000 there was an urgent need to tackle complicated optimization problems. In this period evolutionary-based metaheuristics (evolutionary strategies, evolutionary programming, genetic algorithms, genetic programming, and differential evolution, as the most well-known ones) and trajectory-based metaheuristics (such as hill climbing, simulated annealing, tabu search, iterated local search, variable neighborhood search, as the most well-known ones) were developed.

The transitional period took place around 2000, when the most successful and well-known swarm-based metaheuristics (Particle Swarm Optimization and Ant Colony Optimization) were invented. In this period more powerful metaheuristics as well as novel frameworks for the local and global search were required.

Since 2000, that can be named as the application period, many practical optimization problems were formulated, modeled, and optimized leading to optimum solutions. In this period a great deal of advanced literature in such fields as metaheuristics for multimodal and multi-objective optimization, parallel metaheuristics, hybrid metaheuristics, constraint handling methods for constrained optimization, metaheuristics for large-scale optimization, metaheuristics for expensive optimization, metaheuristics in synergy, and cloud computing was developed [6]. On the other hand, developing new frameworks to reach a more efficient trade-off between exploration and exploitation was of interest more and more in this period, and as a result, many new nature-based metaheuristics were developed. For example, one can refer to harmony search (HS), Artificial Bee Colony (ABC) algorithm, Big Bang Big Crunch (BB-BC), teaching learning-based optimization (TLBO), imperialist competitive algorithm (ICA), Cuckoo Search (CS), Krill Herd (KH), and Gray Wolf Optimizer (GWO) as the most well-known recent metaheuristics and Nephron Algorithm Optimization (NAO), Salp Swarm Algorithm (SSA), Grasshopper Optimization Algorithm (GOA), Lightning Attachment Procedure Optimization (LAPO), and Collective Decision Optimization Algorithm (CDOA) as the most recent ones based on the knowledge of the authors until today. In addition to serious efforts in all advanced fields of metaheuristics and their applications, the first author and his students have introduced many new frameworks for metaheuristics [7, 8]. Some of these are Charged System Search (CSS) algorithm, Ray Optimization (RO) algorithm, Dolphin Echolocation (DE) Optimization, Colliding Bodies Optimization (CBO) algorithm, Tug-of-War Optimization (TWO) algorithm, Water Evaporation Optimization (WEO) algorithm, Vibrating Particles System (VPS) algorithm, Cyclical Parthenogenesis algorithm (CPA), and Thermal Exchange Optimization (TEO) algorithm.

In a nutshell, metaheuristics have emerged as a very interesting and useful field. As a result, nowadays we can see optimization modules in commercial analysis and design software products and solutions. In addition, a number of metaheuristic-based software support systems have been developed to provide customizable tools to solve optimization problems.

1.4 Application Examples of Metaheuristics

Noteworthy improvements in the performance of computers following in the last decades have produced the ability to model, analyze, and design real-world problems as precise as possible in different application and research domains. Almost all real-world problems can be considered as intricate optimization problems. This complexity arises from different inherent characteristics such as discrete or mixed discrete and continuous solution space, highly nonlinear and/or multi-criteria performance evaluation of the problem, large search space of potential solutions, and numerous nonconvex design constraints. On the other hand, emerging significant merits of the metaheuristics make them a prime choice that has achieved increasing interest in optimizing real-world optimization problems. Metaheuristics are used in optimization problems in various disciplines, e.g., engineering, economics, and sciences. It should be noted that metaheuristics have also found applications in solving many combinatorial optimization problems, nondeterministic polynomial-time solvable problems, and various search problems such as feature selection, automatic clustering, and machine learning. In engineering domain one can refer to the following applications: optimum design of aircraft and space vehicles in aerospace engineering; optimization of software cost and time estimation in software engineering; network and computer security in computer engineering; optimum design of structures and infrastructures in civil engineering; optimum design of mechanical, automotive, and robotics systems; mechanical components, devices, tools, and equipment and mechanical processes in mechanical engineering; process optimization in chemical engineering; optimal production planning and scheduling in industrial and system engineering; signal and image processing; and optimum design of networks in electrical engineering and many other applications in various interdisciplinary fields of engineering. The areas of application of metaheuristics have grown to such an extent that nowadays many conferences or sessions in conferences as well as archived journals are being dedicated to metaheuristics and their applications.

References

1. Beightler CS, Phillips DT, Wilde DJ (1967) Foundations of optimization. Prentice-Hall, Englewood Cliffs, NJ
2. Rao SS (2009) Engineering optimization: theory and practice, 4th edn. Wiley, Hoboken, NJ
3. Glover F (1986) Future paths for integer programming and links to artificial intelligence. Comput Oper Res 13:533–549
4. Osman IH, Laporte G (1996) Metaheuristics: a bibliography. Ann Oper Res 63:513–623
5. Sorensen K, Sevaux M, Glover F (2017) A history of metaheuristics. arXiv preprint arXiv, 1704.00853
6. Xiong N, Molina D, Ortiz ML, Herrera F (2015) A walk into metaheuristics for engineering optimization: principles, methods and recent trends. Int J Comput Intell Syst 8(4):606–636

7. Kaveh A (2017) Advances in metaheuristic algorithms for optimal design of structures, 2nd edn. Springer, Cham
8. Kaveh A (2017) Applications of metaheuristic optimization algorithms in civil engineering. Springer, Cham

Chapter 2
Preliminaries and Frameworks

2.1 Introduction

This chapter provides the required preliminaries and presents the needed frameworks to understand the subsequent chapters. After presenting the most basic notions of MATLAB program as well as definitions and notations from metaheuristics for constrained optimization problems, a simplified standard template for metaheuristics is presented. Then the presentation and evaluation frameworks of the considered algorithms are described.

2.2 Brief Introduction to MATLAB

MATLAB (matrix laboratory) which is the trademark of MathWorks company can be considered as a high-level language. It integrates computation, visualization, and programming in an easy manner to use flexible environment represented in familiar mathematical notation [1]. The use of MATLAB is getting widespread with various types of applications such as algorithm development in both academia and industry. In fact, the basic data element in MATLAB is an array that makes it suitable for matrix or vector formulation. On the other hand, population-based metaheuristics make a series of repetitive operations for modification and replacement on a matrix of candidate solutions or the population of the algorithm. In this regard, MATLAB can be an objective selection to program the population-based metaheuristics.

Some needed basic background to understand the programmed algorithms is provided in this section. For a detailed study of MATLAB, the reader can refer to many MATLAB references and the very useful help of the software [1].

A. Kaveh, T. Bakhshpoori, *Metaheuristics: Outlines, MATLAB Codes and Examples*, https://doi.org/10.1007/978-3-030-04067-3_2

2.2.1 Matrices and Vectors and the Operations

An $r \times c$ matrix can be considered as a rectangular array of numbers with r rows and c columns:

$$M = \begin{bmatrix} 1 & 2 & 3 \\ 4 & 5 & 6 \end{bmatrix}$$

$$\text{In MATLAB}: M = [1, 2, 3; 4, 5, 6];\ size(M) = (r = 2, c = 3)$$

Indexing of matrices is made almost in subscript indexing notation (e. g., $A(2, 1) = 4$) and colon indexing notation (e.g., $A(2, 2 : 3) = [5\ 6]$). MATLAB features all the standard calculator commands with the typical order of arithmetic operations on matrices. It has also many useful matrix functions.

It should be noted that row and column vectors are lists of r and c numbers separated by either commas or semicolons with the size of $1 \times n$ and $m \times 1$, respectively.

$$V1 = [1\ \ 2\ \ 3];\ V2 = \begin{bmatrix} 1 \\ 2 \\ 3 \end{bmatrix}$$

$$\text{In MATLAB}: V1 = [1, 2, 3];\ V2 = [1; 2; 3]$$

2.2.2 Flow Controls

Flow control constructs are the most common and useful control structures in MATLAB. The correct usage of control structures and a proper combination of which is a basic task in programming. Here, we introduce the most prevalent control structures: *if* statements, *for* and *while* loops, and *break.*

if executes a block of statement in the case that its conditional expression is evaluated as true. For the case of false evaluation of the conditional expression, using *else* statements, it is possible to execute another block. Using *elseif* it is possible to execute another block of statements. A general syntax of the *if* statement can be presented as follows:

if first conditional expression
first block of statements
elseif second conditional expression
second block of statements
else

third block of statements
end

For and *while* loops are very similar to each other. The *for* loop executes a block of statements in a predetermined number of times. The *while* loop continues the execution of a block of statements as long as some expressions are true. A general syntax of the *for* and *while* loops can be presented as follows:

for index = start: increment, finish a block of statements **end**	**while** expression a block of statements **end**

Sometimes iterative loops should be stopped when a specific state is satisfied. The execution of the *for* and *while* loops can be stopped by the *break* command. It is also beneficial to use the nested loops by applying a loop inside another loop.

2.2.3 Scripts and Functions

A *script* is the simplest type of program in MATLAB with .m extension. Via scripts, it is possible to write a series of commands and have calls of functions in a plain text file and call or execute it as a unit. In addition to many useful functions available in MATLAB vocabulary, new functions can be defined as well. Functions can be defined in two ways: anonymous functions and function M-files. Anonymous functions are usually appropriate for small frames. The latter which can be written in MATLAB editor the same as a script is very useful to make the programs well-organized. Any section of the program with a specific task can be considered as a reusable and easy to check M-file function. All M-file functions should be structured with the following syntax as the first row and should also be named exactly with the name of the function and saved with extension .m:

Function (List of outputs) = The name of function (List of inputs)

In fact, when the function is called in the main program, it returns a list of outputs after execution of a body of statements on the list of input arguments. It should be noted that functions can be used in a nested form. It should also be noted that functions cannot use the variables in the basic workspace. For this purpose (variables available in the basic workspace and are not given as input variables, calling them in the context of a function) and also other preliminaries of MATLAB, readers are encouraged to study various references easily accessible or the useful help of this software.

According to the aforementioned paragraph, the difference between scripts and M-file functions is that the latter has input and output parameters and in comparison with the scripts that execute a series of commands or function called on the hard-coded variables is more flexible. Therefore, in this book, metaheuristics are programmed as script files composed of a specific series of actions. Any part of algorithms with specific and general tasks will be coded as M-file functions and called anywhere needed in the main body of scripts of metaheuristics.

2.3 Definitions and Notations

Optimization is the best way to find a solution yielding an optimal performance of the problem under the circumstances governing the problem. As it was defined in the first chapter, optimization is a three-step decision-making process. Here we try to define the optimization in a simple manner as a two-step process, namely, modeling of the problem and the optimization method.

The performance of the problem at hand is characterized based on an objective function or many objective functions. In this regard, optimization problems are categorized into single-objective and multi-objective optimization problems. These objective functions can be either optimally minimal or maximal. However, any maximization problem can be converted into the minimization form. Almost all real-world optimization problems have various kinds of restrictions to meet some desirable design criteria named as design constraints. In this respect, optimization problems are considered as unconstrained and constrained. The latter is named as constrained optimization problem (COP). In this book, metaheuristics are presented for single-objective COPs.

The following questions should be asked when facing a single-objective COP. What is the function ($f(X)$) to characterize and evaluate performance of the problem? How many design variables (x_i) or dimensions (n) does the problem have? What are the practical ranges ($x_i^l \leq x_i \leq x_i^u$) to be assigned to the design variables? What are the design equality constraints ($h_k(X) = 0, \quad k = 1, 2, \ldots, K$) as well as inequality constraints ($g_j(X) \leq 0, \quad j = 1, 2, \ldots, J$) based on the desired design criteria? Then the single-objective COP can be stated mathematically as:

$$\begin{aligned}
&\mathit{find}\ X = \{x_1, x_2, \ldots x_n\}, \\
&\mathit{to\ Mimimize}\ f(X), \\
&\mathit{subject\ to:} \\
&x_i^l \leq x_i \leq x_i^u, \quad i = 1, 2, \ldots, n \\
&g_j(X) \leq 0, \quad j = 1, 2, \ldots, l \\
&h_k(X) = 0, \quad k = 1, 2, \ldots, m
\end{aligned} \tag{2.1}$$

According to the above statement, the number of variables is n, the number of equality and inequality constraints are m and l, respectively, the objective function to be minimized is $f(X)$, the jth inequality constraint, and the kth equality constraints are $g_j(X)$ and $h_k(X)$, respectively. Usually, by adding a tolerance value (ε), the equality constraints can easily be converted to the inequality constraints

($h_k(X) - \varepsilon \le 0$). Design variables can be assigned as continuous variables or discrete variables. In this regard, a COP is categorized as continuous, discrete, and with mixed design variables. In the case of continuous variables for which the algorithms in this book are presented, the range of the ith variable is $[x_i^l, \ x_i^u]$. It is worthwhile to note that the algorithms can easily be adopted for the discrete case, but it is not guaranteed that an algorithm working well in the continuous case will do for the discrete or mixed case as well. The design variables with the dimension of n build the search space in which each set of design variables (X), named as design vector, is a point in this n-dimensional space. Each design vector can be considered as a candidate solution. Each candidate solution can be either feasible or not feasible. Rarely in the real-world optimization problems can one find unconstrained problem. In an ideal situation, almost all optimization problems are constrained in the realworld. Design constraints divide the search space of the COP into two regions: (1) feasible search space, in which all design constraints are satisfied, and (2) infeasible search space, where one, more, or all design constraints are violated. It should be noted that a massive literature has been developed around constrained optimization techniques that are not within the scope of this book and the interested readers are referred to the specialized literature [2]. However, the most common and simple constraint handling technique is the penalty approach in which using a penalized function, the COP is converted into the unconstrained optimization problem. The penalized function is a summation of the original objective function with the measure of constraint violation multiplied by penalty parameter. There are also many valuable references concerning penalty approach resulting in different penalty functions so as to deal with the inequality, equality, and hybrid constraints. In this book, a simple and conventional formulation of penalty function approach is used:

$$pf(X) = f(X) + \sum_{j=1}^{l+m} \vartheta_j \langle g_j(X) \rangle \tag{2.2}$$

in which $\langle g_j(X) \rangle$ is zero if $g_j(X) \le 0$ whereas $g_j(X)$, otherwise. It should be noted that in this formulation, m equality constraints are considered as converted to the inequality constraints. $\vartheta_j \ge 0$ is the penalty parameter corresponding to the jth inequality constraint, which should be large enough depending on the desired solution equality. Searching of the algorithm may pull some dimensions of the candidate solution outside of the search space or the practical ranges of design variables. There are different techniques to handle this problem. The simplest technique by which the variables outside the range reinstated on the lower and upper bounds will be used.

After modeling the optimization problem, the optimization method can be considered decisive. Asking the following questions is requisite to show the importance of this step. How smooth are the objective function and the constraints? Computationally how expensive it is to evaluate the COP performance in a candidate solution, based on the considered objective function? How large is the dimension of the optimization problem and as a result how large is its search space? What is the nature of the design variables, namely, continuous, discrete, or mixed variables? With increasing growth of computers, it is possible and more desired to handle accurate but time-consuming analysis on the physical systems. In many COPs the constraints and the objective function are not smooth or are not linear and convex. We need to find the feasible optimal solution, and usually, this solution lies on the border of feasible and infeasible regions. Real-world COPs mostly have either discrete or mixed design variable nature. Answering these questions, it can be concluded that many real-world COPs are complex problems. According to the previous chapter, metaheuristics can be suitable for solving complex COPs. On the other hand, a widespread optimization technique cannot be considered to be capable of solving all problems efficiently. Consequently, researchers are intended to develop algorithms as efficient as possible.

An efficient algorithm is the one that is capable of finding solution as faster and better as possible. In other words, an efficient algorithm is an algorithm with high convergence speed and with high accuracy. In other words, an efficient metaheuristic should make a good balance between exploration and exploitation in order not to be trapped in local optima, and has acceptable speed. Metaheuristics manage to actualize this balance with specialized frameworks and formulations along with the randomization. Almost all methods have parameters in the context of formulations to make a more effective balance between exploration and exploitation. The common parameter between all methods is the number of individuals that the population-based metaheuristic works with it. In this book, these parameters will be considered equal to the recommended values in the corresponding literature of each algorithm.

2.4 Standard MATLAB Template for a Metaheuristic

A standard MATLAB template for metaheuristics can be considered as a structure composed of three parts. The first section is the initialization phase in which the algorithm parameters, as well as the design problem, are characterized and defined. Additionally, the first set of candidate solutions is generated and evaluated in this phase. The second section is the main repetitive body of the algorithm in which a new set of candidate solutions is generated from the current set of solutions and is replaced in a repetitive manner until the stopping criterion is fulfilled. There are various stopping criteria, and the most common one, fixed number of algorithm iterations or objective function evaluations, will be utilized. At the third section, the results are monitored.

The procedure of coding the algorithms is in a way that leads to simple codes, so the procedure can easily be traced line by line. It is intended to make use of just primitive and basic functions of MATLAB as long as possible, and it is tried to handle the desired operations via series of simple commands. Making use of characters, strings, logical states, cell arrays, structures, tables, function handles, and so on is avoided.

Initialization

Define properties of the COP: problem dimension or number of design variables (*nd*) and practical range of design variables (*Lb* and *Ub*).

Define the algorithm parameters: for example, the number of algorithm individuals and the maximum number of objective function evaluations (*MaxNFEs*) as the stopping criterion of the algorithm.

Construct the M-file of the objective function (*fobj*) to evaluate the problem performance for a candidate solution. Input arguments which are consist of the newly generated design vector should be evaluated and the practical ranges of the design variables. The function will result in the following as the outputs: corrected design vector if it was outside of the search space, objective function value (*fit*), and penalized objective function value (*pfit*). It can be stated as follows:

[Corrected design vector, *fit*, *pfit*] = *fobj* (Newly generated design vector, *Lb*, *Ub*)

Generate the initial random solutions, and evaluate them calling the *fobj* function defined in the previous step.

Monitor the best candidate solution.

Body of the Algorithm

In this section, the main body of the algorithm which is, in fact, a higher level framework will be structured in a while loop. Considering *NFEs* as the number of function evaluations, the algorithm operated until the current iteration, the expression of the while loop is $NFEs < maxNFEs$. The algorithm performs its iterations until this expression is true which is considered as the simplest and the most common stopping criterion.

Produce the required M-File functions with a specific task for generating and replacing the main components based on the framework and formulations of the algorithm. For this purpose, the input and output arguments should be diagnosed. To make consistency in coding all the algorithms, the input arguments of generating new candidate solution function are considered as follows, but additional input arguments can be required for some algorithms: the matrix of candidate solutions and its corresponding vectors of the objective function (*Fit*) and penalized objective function (*PFit*). The matrix of candidate solutions has the order of the number of algorithm individuals $\times$ *nd*, and *Fit* and *PFit* are row vectors with the order of $1 \times nd$. It should be noted that MATLAB is case-sensitive. For example, *PFit* and *pfit*, as were defined previously, are the different objects. The output of this function will be only the newly generated set of candidate solutions. The input arguments for the replacing function will be the matrix of candidate solutions and its corresponding vectors of the objective function (*Fit*) and penalized objective function (*PFit*), newly generated set of candidate solutions by the generation function, and all the input arguments of the *fobj* function except the first one. More input arguments may also be needed for some algorithms. It is important to mention that the *fobj* function is called in the nested form in the context of replacing function, and each row of the matrix of the newly generated candidate solutions, will be its first input argument. The outputs of replacing functions are the updated matrix of candidate solutions, and its corresponding updated vectors of the objective function and penalized objective function and more outputs also can be resulted in some algorithms. Obviously, the replacing function will be called after the generating function. It should be noted that in some algorithms, there are multiple different frameworks for generation. These frameworks are almost sequential. In these cases, as much as the generation function, the replacing function is also called after that. It should be also noted that in some algorithms, the replacement may be made via a different strategy.

Monitor the best candidate solution.

Save the required results such as the obtained best candidate solution and its corresponding objective function and penalized objective function and the current number of function evaluation (*NFEs*) or the number of iterations of the algorithm (*NITs*). In most algorithms, the iteration number is equal to the number of function evaluation divided by the number of algorithm individuals.

Monitoring the Results
In this section desired results can be monitored using post-processing and visualization on the algorithm performance history in its iterative search process.

2.5 Algorithm Representation

In the following chapters, each chapter is dedicated to one algorithm. To keep a unified manner and consistency in presenting the metaheuristics, firstly a general description and the essence of the inspiration behind them are provided. Then the formulation of the algorithm's framework is presented in detail; the pseudo code and flowchart are derived and described.

The chapter will be continued with the presentation of its MATLAB code. Each algorithm is coded in a simple manner. According to the framework of each algorithm, a certain number of M-file functions are considered for coding the algorithm and named according to their jobs. Initially, these functions are presented with the order of their recall in the algorithm, and at the end, the main body of the algorithm is coded composing three parts: initialization, algorithm cyclic body, and monitoring the results. The required comments are presented with the percent sign (%). For applying the algorithms, reader can simply generate the functions with the same names by coping and pasting the codes. The main body of the algorithm can be saved with any desired name, for example, *Main*. All functions of each algorithm should be saved in a specific path, and the algorithm should be applied by running the *Main* function.

At the end, each chapter will be concluded using the experimental evaluation of the algorithm.

2.6 Experimental Evaluation of the Algorithms

Efficiency in evaluation of the metaheuristics in solving single-objective COPs is important and can be done theoretically or analytically and experimentally. Theoretically, it is questioned that how complex is the algorithm to analyze and how sophisticated the formulation is and also how many parameters are used or, in other words, how the complexity of its space is. Experimentally, researchers are interested to evaluate the algorithm performance in comparison with other algorithms on instances of a specific problem, benchmarks, or problems arising in practice. There are also different criteria to experimentally measure the efficiency such as accuracy, speed, and robustness. Generally speaking, almost all real-world COPs, even in their simplest instances, are timely expensive to be evaluated in a candidate solution. It is validly referable to say that in the optimization process of a real-world COP, the major time and space complexity arises due to the evaluations of the problem. In this regard, many researchers are interested in the theoretically complex metaheuristics without any serious concern about complexity, and with this approach, a massive literature is always developed for any newly generated metaheuristic in order to improve its performance. However, an algorithm with better performance and at the same time theoretically simple would be preferable.

In this book, we consider the original versions of the algorithms. The aim is to characterize the properties of the algorithm. That is why a simple benchmark engineering design problem, a tension/compression spring design problem, with three design variables ($nd = 3$) and four inequality constraints ($l = 4$) is considered to be solved by the coded algorithms. The problem will be solved by algorithms with essentially the same framework: the initial solutions are similarly built, and a similar random series is used. It is important to note again that the experimental evaluation does not make comparison between algorithms: the results for the problem are shown only to illustrate the local and global search properties of algorithms and the influence of the parameters of the algorithms.

Here the M-file of the objective function is coded and will be called in MATLAB codes of the algorithms in the next chapters. The tension/compression spring design problem is a well-known engineering optimization problem and has been a subjective selection for evaluating the performance of newly generated metaheuristics. In Fig. 2.1, this spring is schematized, which should be designed for a minimum weight subjected to constraints on shear stress, surge frequency, and minimum deflection. The design variables are the mean coil diameter D, the wire diameter d, and the number of active coils N which are denoted by x_1, x_2, and x_3, respectively. The problem can be stated with the following cost function:

$$f(x) = (x_3 + 2)x_2 x_1^2 \tag{2.3}$$

to be minimized and constrained with:

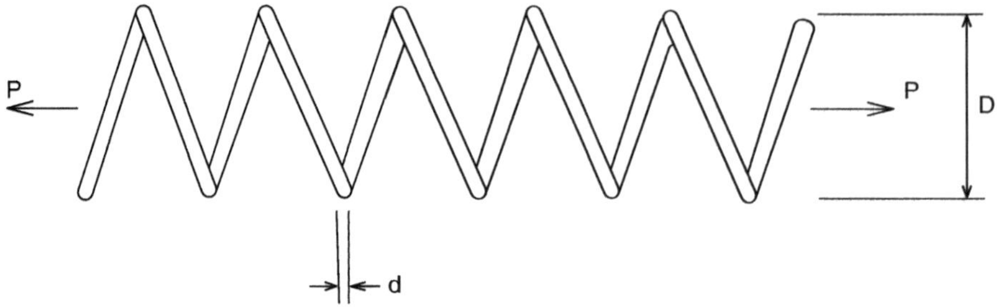

Fig. 2.1 Schematic of the tension/compression spring

$$
\begin{aligned}
g_1(x) &= 1 - \frac{x_2^3 x_3}{71785 x_1^4} \le 0 \\
g_2(x) &= \frac{4x_2^2 - x_1 x_2}{12566\left(x_2 x_1^3 - x_1^4\right)} + \frac{1}{5108 x_1^2} - 1 \le 0 \\
g_3(x) &= 1 - \frac{140.45 x_1}{x_2^2 x_3} \le 0 \\
g_4(x) &= \frac{x_1 + x_2}{1.5} - 1 \le 0
\end{aligned}
\tag{2.4}
$$

The design space is bounded by $0.05 \le x_1 \le 2$, $0.25 \le x_2 \le 1.3$, and $2 \le x_3 \le 15$. The ranges of design variables are $Lb = [0.05\ 0.25\ 2]$ and $Ub = [2\ 1.3\ 15]$.

Based on the definitions for modeling of a COP and penalty approach, presented in the Sect. 2.3, and descriptions of *fobj* function presented in the Sect. 2.4, the objective function is coded as the following:

```
function [X,fit,pfit]=fobj(X,Lb,Ub)

% Correcting the design vector if it is not within the defined range.
for i=1:size(X,2)
    if X(i)>Ub(i)
        X(i)=Ub(i);
    end
    if X(i)<Lb(i)
        X(i)=Lb(i);
    end
end

% Calculate inequality constraints (g(i)). Number of inequality
constraints(l) is 4.
g(1)=1-(X(2)^3*X(3))/(71785*X(1)^4);
g(2)=(4*X(2)^2-X(1)*X(2))/(12566*(X(2)*X(1)^3-X(1)^4))+1/(5108*X(1)^2)-1;
g(3)=1-(140.45*X(1))/(X(2)^2*X(3));
g(4)=(X(1)+X(2))/(1.5)-1;

% Calculate the cost function (fit).
fit=(X(3)+2)*X(2)*X(1)^2;

% Defining the penalty parameter (represented here with nou). Notice that
penalty parameter is considered as a very big number and equal for all four
inequality constraints.
nou=10^9;
penalty=0;
for i=1:size(g,2)
    if g(i)>0
        penalty=penalty+nou*g(i);
    end
end
% Calculate the penalized cost function (pfit) by adding measure of penalty
function (penalty).
pfit=fit+penalty;
```

References

1. Lyshevski SE (2005) Engineering and scientific computations using MATLAB. Wiley, Hoboken
2. Kramer O (2010) A review of constraint-handling techniques for evolution strategies. Appl Comput Intell Soft Comput 2010:185063. 11 pages

Chapter 3
Artificial Bee Colony Algorithm

3.1 Introduction

Swarm intelligence and group behavior of honey bees was the basic inspiration of some metaheuristics. The first one is the Artificial Bee Colony (ABC) algorithm which was introduced by Karaboga in 2005 [1] based on the foraging behavior of honey bees. Other algorithms such as bee colony optimization [2] and bees algorithm [3] were also developed which are not the purpose of this chapter.

In ABC algorithm each candidate solution is represented by a food source, and its nectar quality represents the objective function of that solution. These food sources are modified by honey bees in a repetitive process manner with the aim of reaching food sources with better nectar. In ABC honey bees are categorized into three types: employed or recruited, onlooker, and scout bees with different tasks in the colony. Bees perform modification with different strategies according to their task. Employed bees try to modify the food sources and share their information with onlooker bees. Onlooker bees select a food source based on the information from employed bees and attempt to modify it. Scout bees perform merely random search in the vicinity of the hive. Hence ABC algorithm searches in three different sequential phases in each iteration.

After randomly generating initial bees, iterative process of the algorithm starts until stopping criterion is achieved. Each iteration is composed of three sequential phases. In the first phase which is known as employed or recruited phase, bees search for new food sources based on the information of the individual understandings. In the second phase or onlooker phase, all employed bees share their information of food sources (position and nectar quality) with onlookers in the dance area. The most promising food source is selected by the onlookers based on a selection probability scheme such as the fitness proportionate selection scheme. More onlookers get attracted toward superlative food sources. It should be noted that the number of onlookers is the same as the employed bees and both are the same as the number of food sources around the hive. In other words, every bee whether employee or

A. Kaveh, T. Bakhshpoori, *Metaheuristics: Outlines, MATLAB Codes and Examples*, https://doi.org/10.1007/978-3-030-04067-3_3

onlooker corresponds to one food source. The third and last phase or the scout bee phase starts if a food source cannot be further improved for a predefined number of trials. In this phase, the food source had to be deserted, and its coupled employed bee transformed into a scout bee. The abandoned food sources are replaced with the randomly generated new ones by the scout bees in the search space.

3.2 Formulation and Framework of the Artificial Bee Colony Algorithm

Looking at the foraging behavior of honey bees, there is obviously a good analogy between a population-based metaheuristic and this self-organized biological system. Search space of the problem is all food sources around the hive. Bees, whether employee or onlooker or scout, which search for promising food sources, are the candidate solutions, and the nectar quality of each food source represents the objective function value.

ABC starts with a randomly generated initial honey bees. The number of honey bees (*nHB*) is the first parameter of the algorithm as the size of the colony or population. In the cyclic body of the algorithm, three sequential search phases are performed by employed, onlooker, and scout bees. After search process of each type of bees or generation of new food sources, replacement strategy is performed to keep the old food sources or replace them with newly generated ones. Different replacement strategies can be used in solving COPs. The simplest greedy strategy is used in this book for ABC so that the solution with better quality or smallest *PFit* is preferred to the old one. It should be noted that evaluation of new food sources firstly should be performed in replacement strategy; hence the number of objective function evaluations (*NFEs*) will be updated. In the following, these three phases are formulated. Worth to mention that the scout bee phase is merely for generating and adding new food sources to the population regardless of the quality, then the replacement strategy will not be used in this phase.

1. Generation of new honey bees (*newHB*) based on the recruited or employed bees strategy. Each employed bee attempts to find a new better food source by searching around its corresponding food source with a random permutation-based step size toward a randomly selected other food source except for herself. This phase can be stated mathematically as:

$$\begin{aligned} \text{stepsize} &= rand_{(i)(j)} \cdot (HB - HB[permute(i)(j)]) \\ newHB &= HB + \text{stepsize} \end{aligned} \tag{3.1}$$

where $rand_{(i)(j)}$ is a random number chosen from the continuous uniform distribution on the $[-1, 1]$ interval, *permute* is different rows permutation functions, *i* is the number of honey bees, and *j* is the number of dimensions of the problem. This phase enables the ABC in the aspect of diversification so that each bee attempts to search its own neighborhood. It should be noted that this search takes place over large steps at the beginning of the algorithm and gradually it gets smaller as the population approaches each other with the completion of the algorithm process.

2. Generate new honey bees (*newHB*) based on the onlooker bees strategy. After completing search process of all the employed bees, share their information (nectar quality and position) of corresponding food sources with onlooker bees. The numbers of employed and onlooker bees are the same. Each onlooker bee is attracted by an employed bee with the probability P_i, and she selects a food source associated with that employed bee to generate new food source for possible modification. It seems that onlooker bees are more attracted with food sources with better nectar quality. A selection probability scheme such as the fitness proportionate selection or roulette wheel selection scheme is used in the ABC calculated by the following expression:

$$P_i = PFit_i / \sum_{i=1}^{nHB} PFit_i \tag{3.2}$$

in which $PFit_i$ is the penalized objective function of the *i*th food source. After choosing a food source (HB_{rws}) based on the roulette wheel selection scheme by the *i*th onlooker bee, a neighborhood source is determined by adding a permutation-based random step wise toward a randomly selected food source except herself:

$$\begin{aligned} \text{stepsize} &= rand_{(i)(j)} \cdot (HB_{rws} - HB[permute(i)(j)]) \\ newHB &= \begin{cases} HB_{rws} + \text{stepsize}, & \text{if } rand < mr \\ HB_{rws}, & \text{otherwise} \end{cases} \end{aligned} \tag{3.3}$$

where $rand_{(i)(j)}$ is a random number chosen from the continuous uniform distribution on the $[-1, 1]$ interval, *permute* is different rows permutation functions, *i* is the number of honey bees, and *j* is the number of dimensions of the problem. Another parameter, modification rate (*mr*), is defined in the version of the ABC algorithm for constrained optimization as a control parameter that controls whether the selected food source by onlooker bee will be modified or not. *Rand* is a randomly chosen real number in the range [0, 1]. This parameter is considered here as 0.8 based on the specialized application literature of the algorithm. This phase ensures the intensification capability of the algorithm so that onlooker bees prefer further to explore the neighborhood of the superlative food sources.

3. In scout bee phase, employed bees who cannot modify their food sources after a specified number of trials (*A*) become scouts. The corresponding food source will be abandoned, and a random-based new food source will be generated in the vicinity of the hive.

This phase merely produces diversification and allows to have new and probability infeasible candidate solutions. It sounds that this phase will be active in the

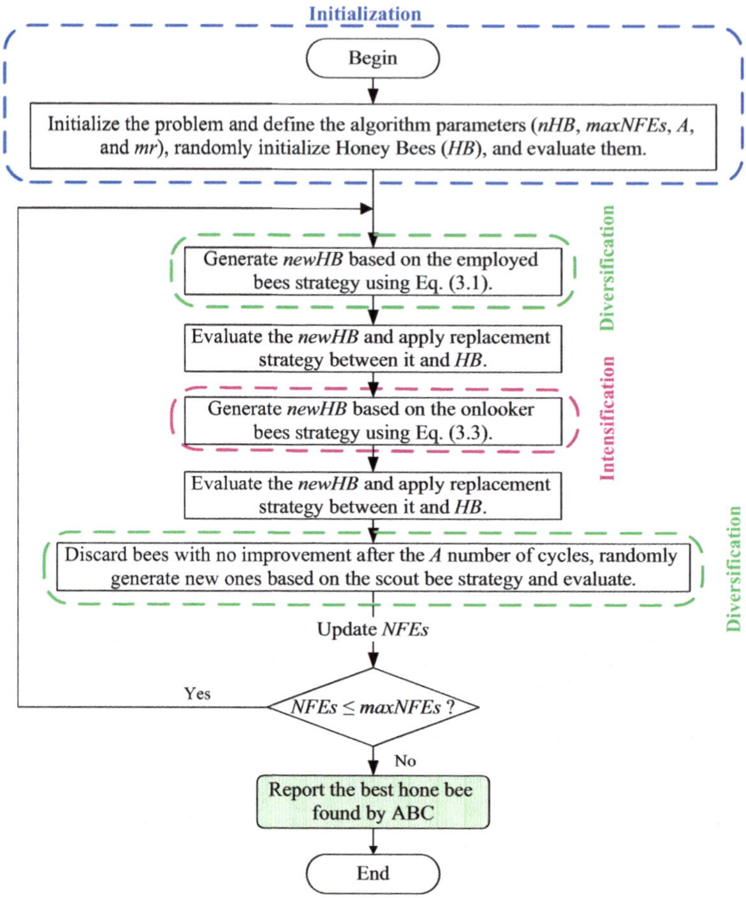

Fig. 3.1 Flowchart of the ABC algorithm

near to end cyclic process of the algorithm. This parameter is considered here as 400 based on the specialized application literature of the algorithm.

The pseudo code of algorithm is provided as follows, and the flowchart of ABC is illustrated in Fig. 3.1.

The pseudo code of the ABC algorithm for solving COPs:

Define the algorithm parameters: *nHB*, *maxNFEs*, *A*, and *mr*.
Generate random initial solutions or food sources (*HB*).
Evaluate initial population or the nectar of initial food sources.
While *NFEs*<*maxNFEs*

Generate new food sources based on the employed bees strategy using Eq. (3.1).
Evaluate the *newHB* and apply replacement strategy between old and new food sources.
Update *NFEs*.

Generate new food sources based on the onlooker bees strategy using Eq. (3.3).
Evaluate the *newHB* and apply replacement strategy between old and new food sources.
Update *NFEs*.
Discard each food source if there is no improvement after the *A* number of cycles. Employ scout bees to generate randomly food sources in the vicinity of the hive and then evaluate them.
Update *NFEs*.
Monitor the best food source.

end While

3.3 MATLAB Code for Artificial Bee Colony Algorithm

The algorithm is coded here in a simple manner. According to the previous section, four functions are considered for coding the algorithm. First, these functions are presented with the order of their recall in the algorithm, and at the end, the algorithm is coded. The required comments are presented. Note that the percent sign (%) represents a comment.

The first function is the objective function evaluation (*fobj*) which was presented in Chap. 2.

The second function is the new honey bee generation function based on the employed or recruited bees strategy. This function is named as *Employed_Bees*. The input and output arguments of this function are the current set of honey bees (*HB*) and the newly generated ones (*newHB*). It should be noted that for getting the permutation-based random step walk, a random bee should be selected except herself. The *Employed_Bees* function is coded as follows.

```
% Generate new Honey Bees (newHB) by the Recruited or Employed Bees
strategy.

function newHB=Employed_Bees(HB)

phi=unifrnd(-1,+1,size(HB));
rp=randperm(size(HB,1));
for i=1:size(rp,2)% Choose a randomly honey bee except herself.
    while rp(i)==i
        rp(i)=round(1+(size(HB,1)-1).*rand);
    end
end
stepsize=phi.*(HB-HB(rp,:));

newHB=HB+stepsize;
```

The third function is the *Replacement* function in which the old bees are compared and replaced with the newly generated ones in a simple greedy manner so that the bee with smaller penalized objective function (*pfit*) will be preferred. The *fobj* function is called in the nested form within this function. The input and output arguments for this function are the same to the *fobj* function but in the matrix form of all candidate solutions. It should be noted that an extra vector argument (*SC*) should be considered for counting the number of no improvements of employed bees to convert them into a scout bee after *A* number of cycles.

```
% Evaluating and Updating Honey Bees by comparing the old and new
ones.

function [HB,Fit,PFit,SC]=Replacemnet(HB,newHB,Fit,PFit,Lb,Ub,SC)

for i=1:size(HB,1),
     [X,fit,pfit]=fobj(newHB(i,:),Lb,Ub);
     if pfit<=PFit(i)
         HB(i,:)=X;
         Fit(i)=fit;
         PFit(i)=pfit;
     else
         SC(i)=SC(i)+1;% Update the counter of not improvement in
the quality of solution.
     end
end
```

The fourth function is the new honey bee generation function based on the onlooker bees strategy. This function is named as *Onlooker_Bees*. The input and output arguments of this function are identical to the *Employed_Bees* function with two additional input arguments: the vector of the penalized objective function (*PFit*) for calculating the selection probability P_i used in the fitness proportionate selection scheme and the modification rate parameter (*mr*). It should be noted that for getting the permutation-based random step walk, a random bee should be selected except herself. The *Onlooker_Bees* function is coded as follows:

```
% Generate new Honey Bees based on the Onlooker Bees strategy.

function newHB=Onlooker_Bees(HB,PFit,mr)

% Calculating the selection probability.
P=(PFit)/sum(PFit);

% Fitness proportionate selection or roulette wheel selection.
for i=1:size(HB,1)
    C=cumsum(P);
    rws(i)=find(rand<=C,1);
end

phi=unifrnd(-1,+1,size(HB));
rp=randperm(size(HB,1));
for i=1:size(rp,2)
    while rp(i)==i
        rp(i)=round(1+(size(HB,1)-1).*rand);
    end
end
stepsize=phi.*(HB(rws,:)-HB(rp,:));

MR=rand(size(HB))>mr;
newHB=HB(rws,:)+stepsize.*MR;
```

The algorithm is coded in the following composed of three parts: initialization, algorithm cyclic body, and monitoring the results.

```
clc
clear

%% Initialization

% Define the properties of COP (tension/compression spring design
problem).
nV=3; % Number of design variables.
Lb=[0.05 0.25 2]; % Lower bounds of design variables.
Ub=[2 1.3 15]; % Upper bounds of design variables.

% Define parameters of ABC algorithm.
nHB=50; % Number of Honey Bees.
maxNFEs=20000; % Maximum number of Objective Function Evaluations.
SC=zeros(nHB,1); % Vector for counting the number of no improvements
to be a Scout Bee.
A=400; % Predetermined number of trials for abandonment.
mr=0.8; % Modification rate.

% Generate random initial solutions.
for i=1:nHB
    HB(i,:)=Lb+(Ub-Lb).*rand(1,nV); % Honey Bees matrix or matrix of the
initial candidate solutions or the initial population.
end

% Evaluate initial population (HB) calling the fobj function constructed in
the second chapter and form its corresponding vectors of objective function
(Fit) and penalized objective function (PFit). It should be noted that the
design vectors all are inside the search space.
for i=1:size(HB,1)
    [X,fit,pfit]=fobj(HB(i,:),Lb,Ub);
    HB(i,:)=X;
    Fit(i)=fit;
    PFit(i)=pfit;
end

% Monitor the best candidate solution (bestHB) and its corresponding
penalized objective function (minPFit) and objective function (minFit).
[minPFit,m]=min(PFit);
minFit=Fit(m);
bestHB=HB(m,:);

%% Algorithm Body

NFEs=0; % Current number of Objective Function Evaluations used by the
algorithm until yet.
NITs=0; % Number of algorithm iterations

while NFEs<maxNFEs

    NITs=NITs+1; % Update the number of algorithm iterations.

    % Generate new Honey Bees (newHB) based on the Recruited Bees strategy.
    newHB=Employed_Bees(HB);

    % Replace the old Bees with the newly generated ones if the newest ones
are beter. Notice that the fobj function is called in the following
replacement function in nested form. Hence the newly generated Bees will be
corrected and evaluated.
    [HB,Fit,PFit,SC]=Replacemnet(HB,newHB,Fit,PFit,Lb,Ub,SC);
```

```
    % Update the number of Objective Function Evaluations used by the
algorithm untill yet.
    NFEs=NFEs+nHB;

    % Generate new Honey Bees (newHB) based on the Onlooker Bees strategy.
    newHB=Onlooker_Bees(HB,PFit,mr);

    % Replace the old Bees with the newly generated ones if the newest ones
are better. Notice that the fobj function will be called in replacement
function in nested form. Hence the newly generated Bees will be corrected
and evaluated.
    [HB,Fit,PFit,SC]=Replacemnet(HB,newHB,Fit,PFit,Lb,Ub,SC);

    % Update the number of Objective Function Evaluations used by the
algorithm until yet.
    NFEs=NFEs+nHB;

    % Transform the Bee coupled with the incapable food source into a scout
bee. If this happens, the transformed Bee should be evaluated. Hence the
fobj function will be called.
    for i=1:nHB
        if SC(i)>=A
            SC(i)=0;
            HB(i,:)=Lb+(Ub-Lb).*rand(1,nV);
            [HB(i,:),Fit(i),PFit(i)]=fobj(HB(i,:),Lb,Ub);
            NFEs=NFEs+1;  % Update the number of Objective Function
Evaluations used by the algorithm until yet.
        end
    end

    % Monitor the best candidate solution (bestHB) and its corresponding
penalized objective function (minPFit) and objective function (minFit).
    [minPFit,m]=min(PFit);
    minFit=Fit(m);
    bestHB=HB(m,:);

    % Display desired information of the iteration.
    disp(['NITs= ' num2str(NITs) '; minFit = ' num2str(minFit) '; minPFit
= ' num2str(minPFit)]);

    % Save the required results for post processing and visualization of
algorithm performance.
    output1(NITs,:)=[minFit,minPFit,NFEs];
    output2(NITs,:)=[min(PFit),max(PFit),mean(PFit)];
    output3(NITs,:)=[bestHB,NFEs];
end

%% Monitoring the results.
figure;
plot((1:1:NITs),output2(:,1),'g',(1:1:NITs),output2(:,2),'r--
',(1:1:NITs),output2(:,3),'b-.')
legend('min','max','mean');
xlabel('NITs');
ylabel('pfit');
axis([1 NITs 0.95*min(PFit) 1.2*min(PFit)])
```

3.4 Experimental Evaluation

As it was mentioned in the second chapter, the aim of experimental evaluation is to show the effects of algorithm parameters on its performance and study the exploration and exploitation capabilities of the algorithm. ABC has four parameters consisting of the maximum number of objective function evaluations (*maxNFEs*) as the stopping criterion of the algorithm, number of honey bees (*nHB*), specified number of trials (*A*) by which an employed bee becomes a scout bee, and the modification rate (*mr*) as a control parameter that controls whether the selected food source by onlooker bee will be modified or not.

Considering a large enough value for *maxNFEs* to ensure convergence of the algorithms is essential. This parameter is almost problem dependent and should be considered larger for complicated and larger problems. It is considered equal to 20,000 for ABC and other algorithms in the next chapters.

The number of individual algorithm in the population-based metaheuristics can be a very important parameter that influences the performance of the algorithm. Figure 3.2 shows the convergence history for a single run of ABC algorithm and identical initial population with different values of *nHB* from 5 to 200. It should be noted that partners of A and *mr* are considered as 400 and 0.8, respectively, in these runs. According to this figure, the values of *nHB* between 20 and 100 are sufficient, and numbers of 50 result better in both aspects of the accuracy and convergence rate. It is worth mentioning that the ABC gets trapped in local optimum with the *nHB* less than 10. On the other hand, large values reduce the convergence rate.

For studying the influence of the employed and onlooker bees strategies on the algorithm performance, Fig. 3.3 monitors convergence history of the ABC together with two other cases: ABC with onlooker and employed bee phases alone.

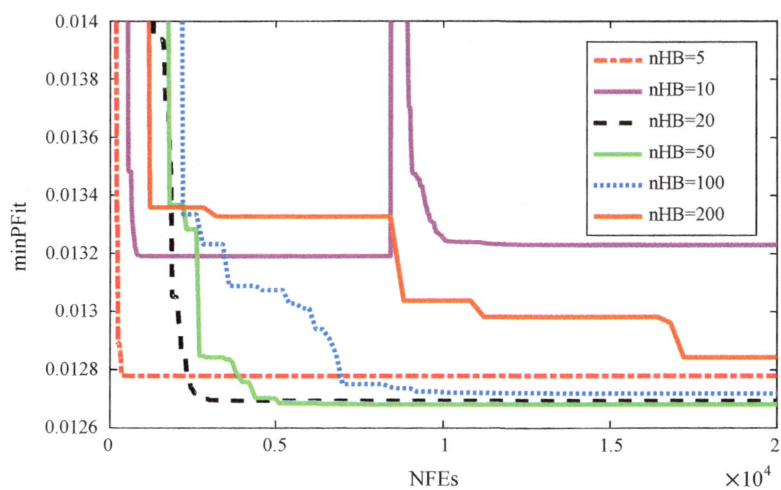

Fig. 3.2 Convergence histories of the algorithm for different values of *nHB*

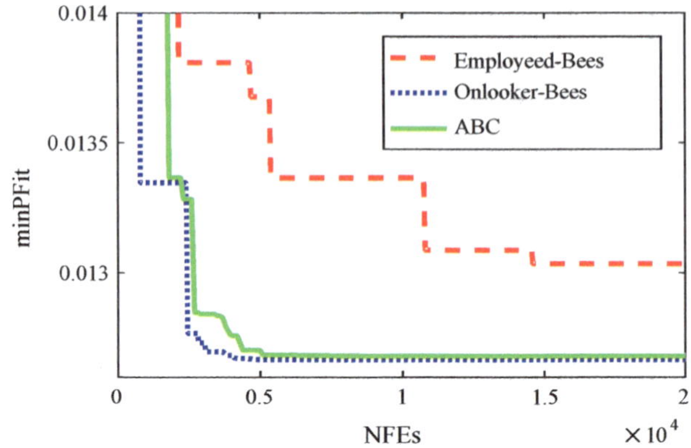

Fig. 3.3 Convergence histories of the ABC together with and without its search strategies

maxNFEs, *nHB*, *A*, and *mr* are considered as 20,000, 50, 400, and 0.8, respectively, and all runs are carried out from a same initial population. For the sake of clarity, the *Y* axis is limited. As it is clear, ABC cannot perform well without considering the onlooker bees strategy which characterizes the intensification of the algorithm. It should be noted that the ABC with only the onlooker bees strategy performs slightly better than the ABC, in both aspects of accuracy and conveyance rate. It should be noted that in both employed and onlooker bees strategies, the random walks are performed based on the random permutation based on the current solutions. In the first iterations, honey bees are farther from each other than last iterations. In this way, the generated permutation-based step size will guarantee somehow the global and local search capability with processing the algorithm iterations. The random part will guarantee the algorithm to be sufficiently dynamic. Then considering onlooker bee phase alone which is the elitism or intensification capability of the algorithm can result in high accuracy and convergence speed. However, it should be noted that in the large and complicated problems, the ABC with both strategies works efficiently and even the necessity of the scout bee phase is clearly determined which is not studied here.

The *A* parameter by which an employed bee becomes a scout bee is considered differently from 40 to 400, and the results are monitored in Fig. 3.4 for a single trial run carried out from the same initial population. This figure shows the convergence history of the best, worst, and mean values of the penalized objective function of population. For better clarity, the *Y* axis presented in the logarithmic scale. Other parameters are considered as $maxNFEs = 20{,}000$, $nHB = 50$, and $mr = 0.8$. Considering small values results in extra diversification so that in the case of $A = 40$, the min, max, and mean diagrams do not converge to each other even in this problem which is known to be a simple and small COP. For additional comparison, Fig. 3.5 depicts the convergence history of best-penalized objective function for these cases. As it is clear, considering $A = 400$ results in a better performance in both aspects of the accuracy and convergence speed. It should be

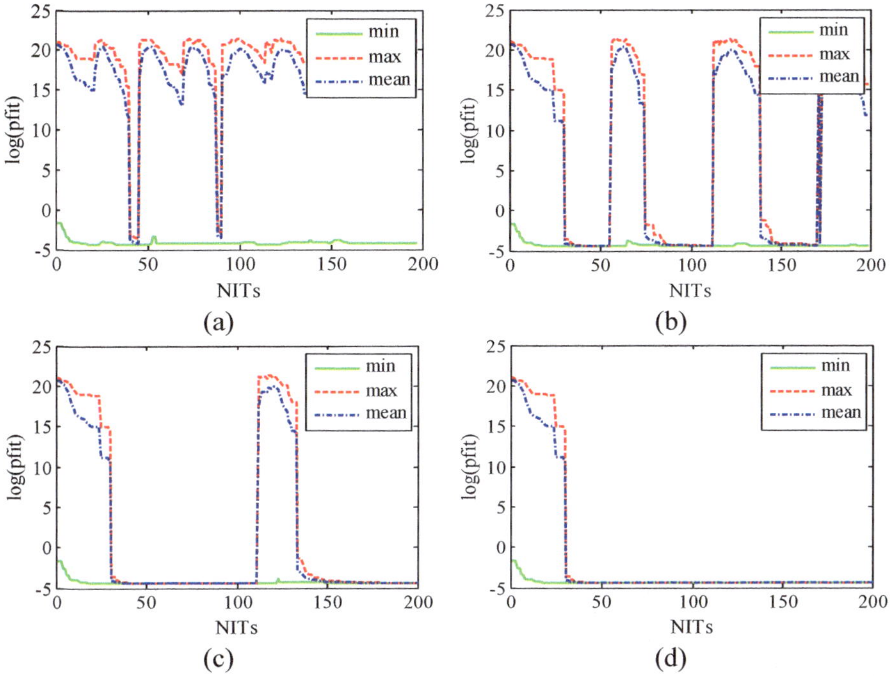

Fig. 3.4 Convergence histories of the minimum, maximum, and mean of the algorithm population for different values of *A* parameter: (**a**) 40, (**b**) 100, (**c**) 200, (**d**) 400

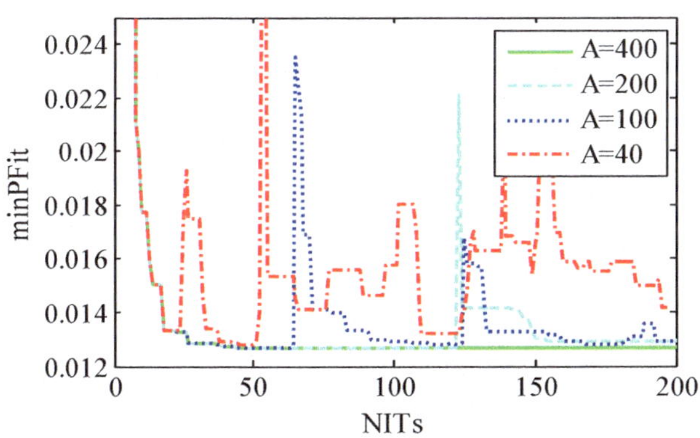

Fig. 3.5 Convergence histories of the ABC with different values of the *A* parameter

noted that for the small values of *A*, the algorithm performs more evaluations of the objective function, in this regard the number of algorithm iterations (*NITs*) is less than 200 in these cases.

Considering $maxNFEs = 20{,}000$, $nHB = 50$, and $A = 400$, Fig. 3.6 shows the convergence histories for a single trial run from the same initial population for

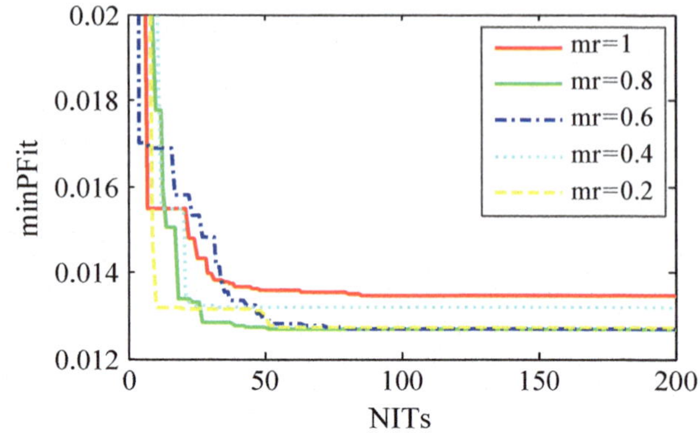

Fig. 3.6 Convergence histories of the ABC with different values of the *mr* parameter

different values of *mr* parameter varying between 0.2 and 1.0. The *mr* value equal to 0.8 results in a better performance of the algorithm. Values less than 0.8 also result in good accuracy and convergence rate, but as it is clear, the dynamic search nature of the algorithm fades, and the step-like movements of the convergence histories disappeared.

References

1. Karaboga D (2005) An idea based on honey bee swarm for numerical optimization. Technical Report, Erciyes University, Engineering Faculty Computer Engineering Department, Erciyes, Turkey
2. Teodorovic D, Dell'Orco M (2005) Bee colony optimization, a cooperative learning approach to complex transportation problems. In: Proceedings of the 10th meeting of the EURO working group on transportation, Poznan, Poland, September, pp 51–60
3. Pham DT, Kog E, Ghanbarzadeh A, Otri S, Rahim S, Zaidi M (2006) The bees algorithm, a novel tool for complex optimisation problems. In: Proceedings of the 2nd international virtual conference on intelligent production machines and systems (IPROMS), Cardiff, UK, pp 454–459

Chapter 4
Big Bang-Big Crunch Algorithm

4.1 Introduction

Inspired by the energy dissipation in the form of transformation from an ordered state to a disordered or chaiotic state, a novel and simple nature-based or physics-based metaheuristic has been developed by Erol and Eksin [1]. The algorithm named as Big Bang-Big Crunch (BB-BC) is taken from the prevailing evolutionary theory for the origin of universe: the Big Bang Theory. According to this theory, in the Big Bang phase, particles are drawn toward irregularity by losing energy, while in the Big Crunch phase, they converged toward a specific direction. Like other population-based metaheuristics, BB-BC starts with a set of random initial candidate solutions, as the initial Big Bang. In fact, each Big Bang phase is preceded with a Big Crunch phase except the first population which should be generated randomly within the search space. After each Big Bang phase, a Big Crunch phase should take place to determine a convergence operator by which particles will be drawn into an orderly fashion in the subsequent Big Bang phase. The convergence operator can be the weighted average of the positions of the candidate solutions or the position of the best candidate solution. These two contraction (Big Crunch) and dispersing (Big Bang) phases are repeated in the cyclic body of the algorithm in succession to satisfy a stopping criteria with the aim of steering the particles toward the global optimum.

4.2 Formulation and Framework of the Big Bang-Big Crunch Algorithm

According to the Big Bang Theory as the prevailing evolutionary theory for the origin of the universe, energy loss causes a transition from an ordered state to a chaotic state. The Big Bang Theory involves two phases: Big Bang phase produces irregularities and disordering of the particles in terms of energy dissipation and Big

A. Kaveh, T. Bakhshpoori, *Metaheuristics: Outlines, MATLAB Codes and Examples*, https://doi.org/10.1007/978-3-030-04067-3_4

Crunch phase in which the scattered particles are converged toward a specific direction. Looking at the Big Bang Theory, there is an analogy between this physical- or natural- or astronomy-based theory and a population-based metaheuristic. Each particle can be considered as an individual of the algorithm population or candidate solution. A certain number of particles are repetitively updated in the search space as the Big Bang phase with step sizes based on the convergence operators of the Big Crunch phase with the aim of condensing around the global optimum of the problem.

Big Bang-Big Crunch (BB-BC) algorithm starts with a set of randomly generated initial solutions in the search space like other population-based metaheuristics. Each cycle of the algorithm is composed of two phases: first the Big Crunch phase in which a converging operator is formed and second the Big Bang phase by which particles are updated in the search space with the step sizes in the vicinity of the converging operator generated in the first phase. Consider a certain number pf particles (*nP*) as the population or candidate solutions matrix (*P*), its corresponding penalized objective function (*PFit*) vector, and the best observed particle in each iteration (*bestP*) with the smallest value of penalized objective function.

The convergence operator based on the Big Crunch phase can be defined as weighted average of the candidate solution positions known as center of mass (*CM*) or the position of the best candidate solution (*bestP*). For minimization problems, *CM* is formulated as:

$$CM(i) = \sum_{j=1}^{nP} (P(j,i)/PFit(j)) / \sum_{j=1}^{nP} (1/PFit(j)), \quad i = 1, \ldots, nv \tag{4.1}$$

The Big Bang phase now can be occurred. In the original BB-BC [1], particles are updated simply with respect to the previously determined center of mass (*CM*) or the position of the best particle (*bestP*) by displacing by a random fraction of the allowable step size defined by the upper (*Ub*) and lower (*Lb*) bound of design variables:

$$newP = (CM \; or \; bestP) + \frac{rand \times (Ub - Lb)}{nIT} \tag{4.2}$$

where *rand* is a random number uniformly distributed in (0, 1). The step size also is divided by the number of the algorithm iterations or number of the Big Bang phases (*NITs*) to generate the effective search range about global optimum or center of mass with the aim of narrowing the search with the progress of the algorithm. It is obvious that the algorithm has two parameters which are required for all metaheuristics: number of the algorithm population and the maximum number of algorithm iterations as the stopping criteria. Camp [2] presented a new formulation with two additional parameters for the Big Bang phase and showed the efficiency of the newly proposed formulation. Considering a parameter for limiting the size of the search space (α) and defining a parameter (β) for considering both *CM* and *bestP* as the converging operator, the new formulation was proposed as:

$$newP(i) = (\beta \times CM + (1 - \beta) \times bestP) + \frac{rand \times \alpha \times (Ub - Lb)}{nIT}, \quad i = 1, \ldots, nP \tag{4.3}$$

This modified formulation is used and coded here. It should be noted that BB-BC algorithm doesn't need replacement strategy. In the other words, particles leave their position no matter if their current position is better.

The pseudo code of algorithm is provided as follows, and the flowchart of BB-BC is illustrated in Fig. 4.1.

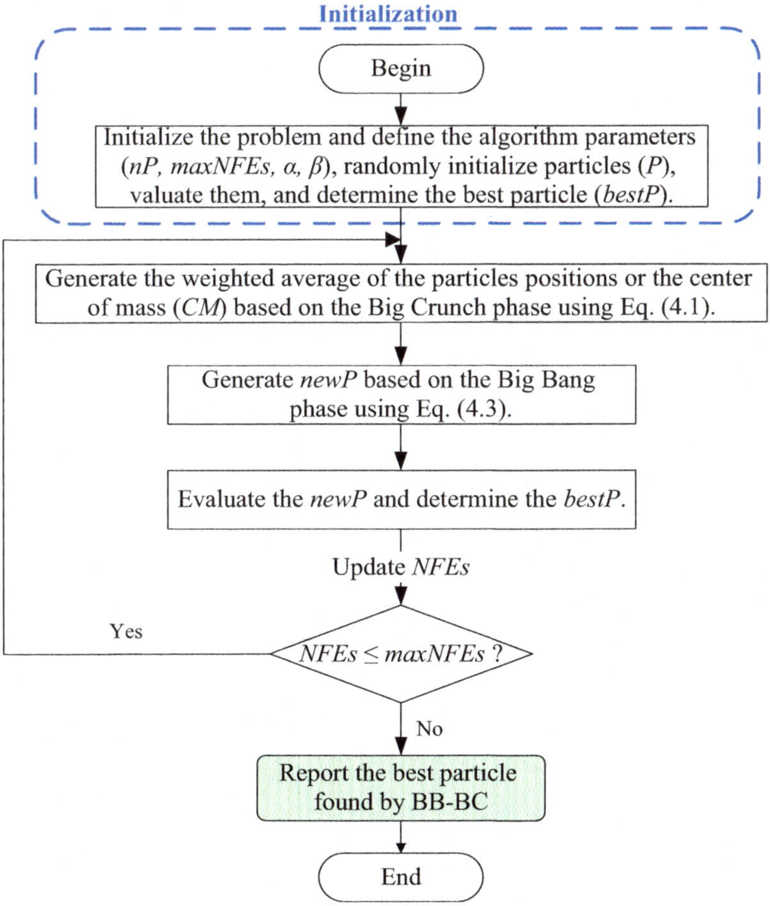

Fig. 4.1 Flowchart of the BB-BC algorithm

The pseudo code of the BB-BC algorithm for solving COPs:

Define the algorithm parameters: *nP*, *maxNFEs*, α, and β.
Generate random initial solutions or particles (*P*).
Evaluate initial population and determine the best particle (*bestP*).
While *NFEs*<*maxNFEs*

Update the number of algorithm iterations (*NITs*).
Generate the weighted average of the particle's position or the center of mass (*CM*) based on the Big Crunch phase using Eq. (4.1).
Generate new particles (*newP*) based on the modified formulation of the Big Bang phase using Eq. (4.3).
Evaluate the *newP* and determine the best particle in this iteration (*bestP*).
Update *NFEs*.
Monitor the best particle found by the algorithm so far.

end While

According to this flowchart, it is obvious that one cannot explicitly distinguish between intensification and diversification because the BB-BC involves these two features together.

4.3 MATLAB Code for Big Bang-Big Crunch Algorithm

The algorithm is coded here in a simple manner. According to the previous subsection, three functions are considered for coding the algorithm. First, these functions are presented with the order of their recall in the algorithm, and at the end, the algorithm is coded. The required comments are presented. Note that the percent sign (%) represents a comment.

The first function is the objective function evaluation (*fobj*) which was presented in Chap. 2.

The second function is the center of mass (*CM*) generation function based on the weighted average of the particle's position and corresponding penalized objective function values. This function is named as *Big_Crunch*. The input arguments are the current set of particle's position (*P*) and corresponding vector of penalized objective function (*PFit*), and the only output argument is the *CM* vector. The *Big_Crunch* function is coded as it follows.

```
% Calculate the Center of Mass (CM) based on the Big Crunch phase.

function CM=Big_Crunch(P,PFit)

for i=1:size(P,2)
    CM(i)=sum(P(:,i)'./PFit)/sum(PFit.^-1);
end
```

The third function is the new particle's generation function based on the Big Bang phase. This function is named as *Big_Bang*. The input arguments of this function are particle's current position (*P*), center of mass vector (*CM*), the current best particle of the population (*bestP*), algorithm-controlling parameters (α and β), number of algorithm iterations (*NITs*), and lower and upper bound of design variables (*Lb* and *Ub*). The output argument is the new position of particle's matrix (*newP*). The *Big_Bang* function coded as it follows:

```
% Generate new set of particles based on the Big_Bang phase

function newP=Big_Bang(P,CM,bestP,beta,alfa,Lb,Ub,NITs)

for i=1:size(P,1)
    newP(i,:)=beta*CM+(1-beta)*bestP+randn(1,size(P,2)).*((Ub-Lb)/(NITs));
end
```

The algorithm is coded in the following composed of three parts: initialization, cyclic body of the algorithm, and monitoring the results. BB-BC uses the best particle of each current iteration named as *bestP*. The values of objective function and penalized objective function of this particle are called as *minFit* and *minPFit*. These values are changed with the algorithm progress and can get worse instead of getting improved. It should be noted that BB-BC attempts to improve the whole population together even if it loses the position of the best particle in the preceding iteration. However, to monitor the algorithm performance, we had to keep in mind the best particle found by BB-BC until *NITs* iteration are called as *BestP* and its corresponding values of objective function (*MinFit*) and penalized objective function (*MinPFit*).

```
clc
clear
%% Initialization

% Define the properties of COP (tension/compression spring design problem).
nV=3; % Number of design variables.
Lb=[0.05 0.25 2]; % Lower bounds of design variables.
Ub=[2 1.3 15]; % Upper bounds of design variables.

% Define parameters of BB-BB algorithm.
nP=50; % Number of Particles.
maxNFEs=20000; % Maximum number of Objective Function Evaluations.
beta=0.2; % Parameter for controlling the influence of the weighted average
of the particles positions or the center of mass (CM) and the best
particle.
alfa=1; % Parameter for limiting the size of the initial search space.

%Generate random initial solutions.
for i=1:nP
    P(i,:)=Lb+(Ub-Lb).*rand(1,nV); %Particles matrix or matrix of the
initial candidate solutions or the initial population.
end

% Evaluate initial population (P) calling the fobj function constructed in
the second chapter and form its corresponding vectors of objective function
(Fit) and penalized objective function (PFit). It should be noted that the
design vectors all are inside the search space.
for i=1:size(P,1)
    [X,fit,pfit]=fobj(P(i,:),Lb,Ub);
    P(i,:)=X;
    Fit(i)=fit;
    PFit(i)=pfit;
end

% Monitor the best candidate solution (bestP) and its corresponding
penalized objective function (minPFit) and objective function (minFit).
[minPFit,m]=min(PFit);
minFit=Fit(m);
bestP=P(m,:);

%% Algorithm Body

NFEs=0; % Current number of Objective Function Evaluations used by the
algorithm until yet.
NITs=0; % Number of algorithm iterations

while NFEs<maxNFEs

    % BestP is the best particle found by the BB-BC until yet. MinPFit and
MinFit are its corresponding penalized objective function and objective
function, respectively.
    if NITs==0
        BestP=bestP;
        MinPFit=minPFit;
        MinFit=minFit;
    end
    NITs=NITs+1; % Update the number of algorithm iterations.
```

```
    % Generate center of mass vector (CM) based on the Big_Crunch phase.
    CM=Big_Crunch(P,PFit);

    % Update the particles position and generate new set of solutions based
on the Big_Bang phase.
    newP=Big_Bang(P,CM,bestP,beta,alfa,Lb,Ub,NITs);

    % Evaluate the new particles. It should be noted that in the BB-BC
algorithm the replacement strategy is not used.
    for i=1:size(P,1)
        [X,fit,pfit]=fobj(newP(i,:),Lb,Ub);
        P(i,:)=X;
        Fit(i)=fit;
        PFit(i)=pfit;
    end

    % Update the number of Objective Function Evaluations used by the
algorithm until yet.
    NFEs=NFEs+nP;

    % Monitor the best candidate solution (bestP) and its corresponding
penalized objective function (minPFit) and objective function (minFit).
    [minPFit,m]=min(PFit);
    minFit=Fit(m);
    bestP=P(m,:);

    if minPFit<=MinPFit
        BestP=bestP;
        MinPFit=minPFit;
        MinFit=minFit;
    end

    % Display desired information of the iteration.
    disp(['NITs= ' num2str(NITs) '; MinFit = ' num2str(MinFit) '; MinPFit
= ' num2str(MinPFit)]);

    % Save the required results for post processing and visualization of
algorithm performance
    output1(NITs,:)=[minFit,minPFit,NFEs];
    output2(NITs,:)=[minPFit,max(PFit),mean(PFit)];
    output4(NITs,:)=[MinFit,MinPFit,NFEs];
    output5(NITs,:)=[BestP,NFEs];
end

%% Monitoring the results
figure(1);
plot((1:1:NITs),output2(:,1),'g',(1:1:NITs),output2(:,2),'r--
',(1:1:NITs),output2(:,3),'b-.')
legend('min','max','mean');
xlabel('NITs');
ylabel('pfit');
figure(2);
plot((1:1:NITs),output4(:,2),'g')
xlabel('NITs');
ylabel('MinFIt');
```

4.4 Experimental Evaluation

The original version of BB-BC has two parameters which are common between all metaheuristics: number of population (*nP*) and maximum number of algorithm iterations (*maxNFEs*). The modified version has extra tow parameters: α and β. Recall that α limits the size of the initial search space and that β determines the contribution of the center of mass and the current best particle in the big bang phase. Actually α and β are equal to 1 in the original BB-BC.

Considering a large enough value for *maxNFEs* to ensure convergence of the algorithms is essential. This parameter is almost problem dependent and should be considered larger for complicated and larger problems. It is considered equal to 20,000 for BB-BC like previous chapters.

The number of algorithm individuals in the population-based metaheuristics can be a very important parameter that influences the performance of the algorithm. Figure 4.2 shows the convergence history for a single run of BB-BC algorithm with identical initial population with different values of *nP* from 5 to 200. It should be noted that α and β parameters are considered as 1 and 0.2, respectively, in these runs. According to this figure, the value of *nP* between 10 and 50 is satisfactory and leads to both suitable accuracy and convergence speed. It is worth mentioning that the BB-BC with *nP* more than 50 performs very poor in terms of accuracy and speed.

For studying the influence of α and β parameters, algorithm is performed for different values of them considering *nP* is equal to 20 from a fixed initial population. Figures 4.3 and 4.4 depict the convergence history of the algorithm for different values of α and β. It should be noted that β and α are considered equal to 0.2 and 1 for these figures, respectively. It should be noted that upper bound of the *Y* axis is limited to more clarity. Anyway, all diagrams start from a joint point. As it is clear, considering small values for β and larger values for α yields better performance of the algorithm. Based on the Fig. 4.4, there is not much difference for the α values more than 0.4. As we know, the value of α in the original BB-BC is equal to 1. Therefore, in this problem, considering the α parameter for limiting the size of

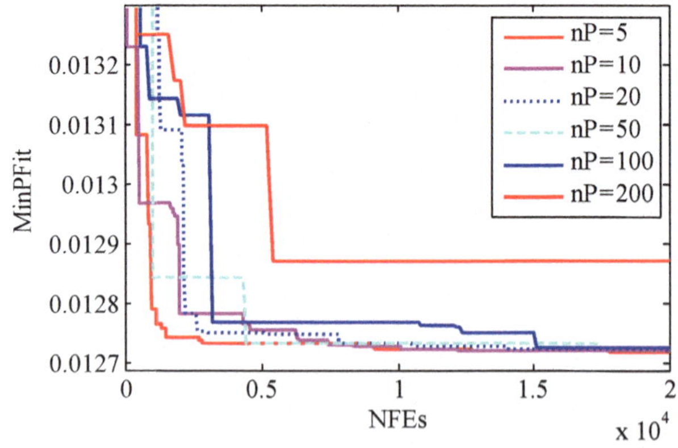

Fig. 4.2 Convergence histories of the BB-BC for different values of *nP*

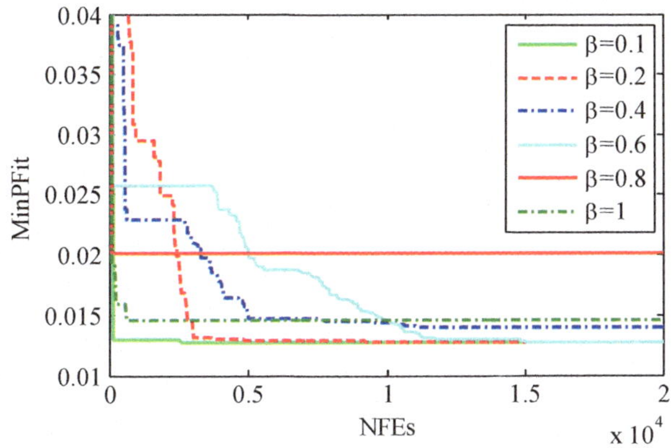

Fig. 4.3 Convergence histories of the BB-BC for different values of β parameter

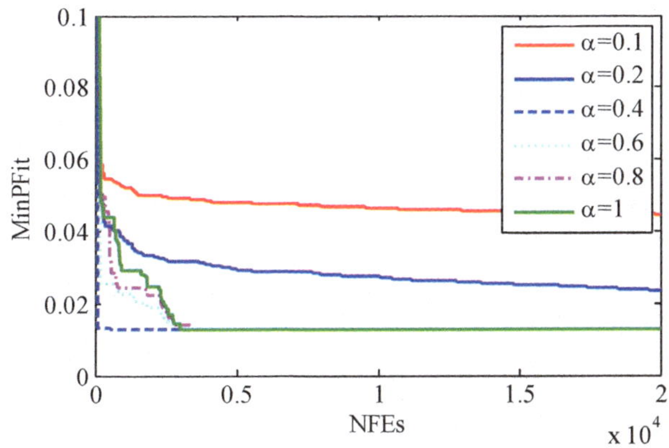

Fig. 4.4 Convergence histories of the BB-BC for different values of α parameter

the initial search space is not requisite, and it is considered equal to 1 in this chapter like the original BB-BC. Such an observation was also apparent by Camp [2] in the optimum design of trusses problem.

In all the previous convergence histories, the *Y* axis monitors the penalized objective function of the best particle (*BestP*) found by the algorithm so far (*MinPFit*). To show how BB-BC attempts to modify the position of the whole population, Fig. 4.5 shows the convergence history of the best, worst, and mean values of the penalized objective function of population. It should be noted that min stands for the penalized objective function of the current best particle (*bestP*). Convergence histories are monitored for constant values of $\alpha = 1$, $\beta = 0.2$, and different values of *nP* and initiated from a fixed initial population. The swinging trend in the convergence histories is due to this fact that BB-BC does not have the replacement strategy and uses the current best particle even if it is worse than the one

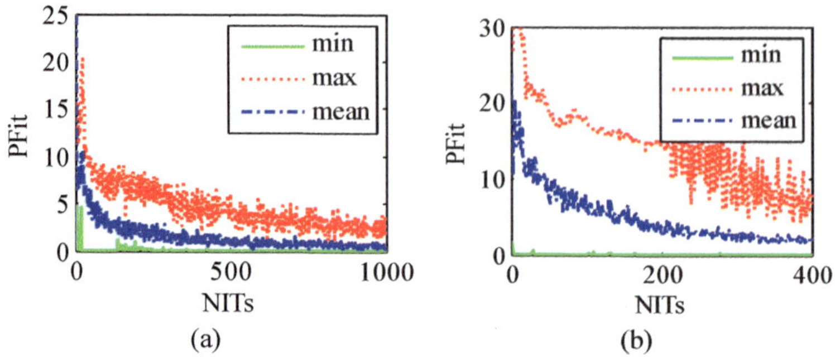

Fig. 4.5 Convergence histories of the minimum, maximum, and mean of the algorithm population for different values of *nP* parameter: (**a**) 20, (**b**) 50

found in the preceding iteration. The more interesting point that should be noted is that the plots in general are descending and BB-BC aims to modify all the population together. However, considering more population results in more diversification.

References

1. Erol Osman K, Eksin I (2006) New optimization method: Big Bang-Big Crunch. Adv Eng Softw 37(2):106–111
2. Camp CV (2007) Design of space trusses using Big Bang–Big Crunch optimization. J Struct Eng 133(7):999–1008

Chapter 5
Teaching-Learning-Based Optimization Algorithm

5.1 Introduction

Teaching-learning-based optimization (TLBO) algorithm was firstly developed in 2011 by Rao et al. [1] based on the classical school learning process. This algorithm consists two stages: effect of a teacher on learners and the influence of learners on each other. TLBO was initialized with a population of random solutions, named students or learners. The best learner or the smartest student with the best objective function is assigned as the teacher in each iteration of the algorithm. Students are updated iteratively to search the optimum within two phases: based on the knowledge transfer from a teacher first (teacher phase) and then from interaction with other students (learner phase). In TLBO the performance of the class in learning or the performance of teacher in teaching is considered as a normal distribution of marks obtained by the students. The main difference of two normal distributions is in their mean value, i.e., a better teacher teaches students with higher average scores. In this regard, TLBO improves other students in the teacher phase by using the difference between the teacher's knowledge and the average knowledge of all the students. The knowledge of each student is obtained based on the position taken place by that student in the search space. In a class, students also improve themselves via interacting with each other after the teaching is completed by the teacher. In the learner phase, TLBO improves each student by the knowledge interaction between that student and another randomly selected one.

A. Kaveh, T. Bakhshpoori, *Metaheuristics: Outlines, MATLAB Codes and Examples*, https://doi.org/10.1007/978-3-030-04067-3_5

5.2 Formulation and Framework of the Teaching-Learning-Based Optimization Algorithm

Looking at the classical school education process, there is obviously a good analogy between a population-based metaheuristic and this social system. Two main components of a teaching and learning process are the influence of a teacher on the students and the interaction accrued between students themselves. In this regard, TLBO considers two sequential phases of search in each of its iterations named as teacher phase and learner phase. Students or learners or classes are the population of the algorithm. The number of learners (*nL*) is the first and the only parameter (except the *maxNFEs* as the stopping criteria parameter) of the TLBO algorithm. Algorithm population is considered as a $nL \times$ number of design variables (nV) matrix denoted by L. Knowledge level of the students is the quality or the value of penalized objective function (*PFit* vector) evaluated based on their position in the search space. The best student is considered as teacher (T).

TLBO starts with randomly generated initial learners. The initial population can be considered as the uneducated students or learners or class. In the cyclic body of the algorithm, two sequential search phases will be performed in each iteration of the algorithm: teacher phase and learner phase. After each searching phase, the replacement strategy is performed to keep the old learners or replace them with newly educated ones. Different replacement strategies can be used in solving COPs. The simplest greedy strategy is used in this book for TLBO so the solution with better quality or smallest *PFit* will be preferred to the old one. In the following, these two phases are presented and formulated:

1. Generation or education of the new learners (*newL*) based on the teacher phase. The class performance as a normal distribution of grades obtained by students can be characterized with the mean value of the distribution. In this phase TLBO aims to improve the class performance by shifting the mean position of the class individuals toward the best learner which is considered as the teacher. This phase is the elitism or global search or intensification ability of the algorithm. In this regard, TLBO updates the learners by a step size toward the teacher obtained based on the difference between the teacher's position and the mean position of all students combining with randomization. Considering the mean position of students in the search space as *MeanL*, this phase can be formulated as follows:

$$\begin{gathered} \text{stepsize}_i = T - TF_i \times MeanL \\ newL = L + rand_{i,j} \times \text{stepsize} \\ i = 1,2,\ldots,nL \text{ and } j = 1,2,\ \ldots,nV \end{gathered} \tag{5.1}$$

in which $rand_{(i)(j)}$ is a random number chosen from the continuous uniform distribution on the [0, 1] interval and TF is a teaching factor considered for controlling how much the teacher will change the mean knowledge of the class which can be either 1 or 2.

2. Generating new learners (*newL*) or updating the knowledge of students by interacting with each other in the learner phase. In this phase, each student interacts with a randomly selected one (L_{rp}) except him or her for possible improvement of knowledge. After comparison, the student will be moved toward the randomly selected one if it is smarter ($PFit_i < PFit_{rp}$) and shifted away otherwise. The learner phase can be stated mathematically in the following equation:

$$\begin{gathered}\text{stepsize}_i = \begin{cases} L_i - L_{rp} & PFit_i < PFit_{rp} \\ L_{rp} - L_i & PFit_i \geq PFit_{rp} \end{cases} \\ newL = L + rand_{i,j} \times \text{stepsize} \\ i = 1,2,\ldots,nL \text{ and } j = 1,2,\ \ldots,nV \end{gathered} \tag{5.2}$$

in which $rand_{(i)(j)}$ is a random number chosen from the continuous uniform distribution on the [0, 1] interval. The learner phase is the local search or diversification capability of the algorithm by which each individual tries to improve by searching its neighborhood and sharing information with one randomly selected individual. It should be noted that the step size of search will be decreased gradually as the students approach each other with the progress of the algorithm.

The pseudo code of TLBO is given as follows, and the flowchart is illustrated in Fig. 5.1.

The pseudo code of the TLBO algorithm for solving COPs:

Define the algorithm parameters: *nL*, *maxNFEs*.
Generate random initial learners or students or the class (*L*).
Evaluate initial learners or the knowledge of initial students.
While *NFEs*<*maxNFEs*

Determine the teacher (*T*) and mean position of all students, and educate or generate new students (*newL*) based on the teacher phase using Eq. (5.1).
Evaluate the *newL* and apply replacement strategy between old and educated class.
Update *NFEs*.
Upgrade or generate new students interacting with each other based on the learner phase using Eq. (5.2).
Evaluate the *newL* and apply replacement strategy between old and learned students.
Update *NFEs*.
Monitor the best student.

end While

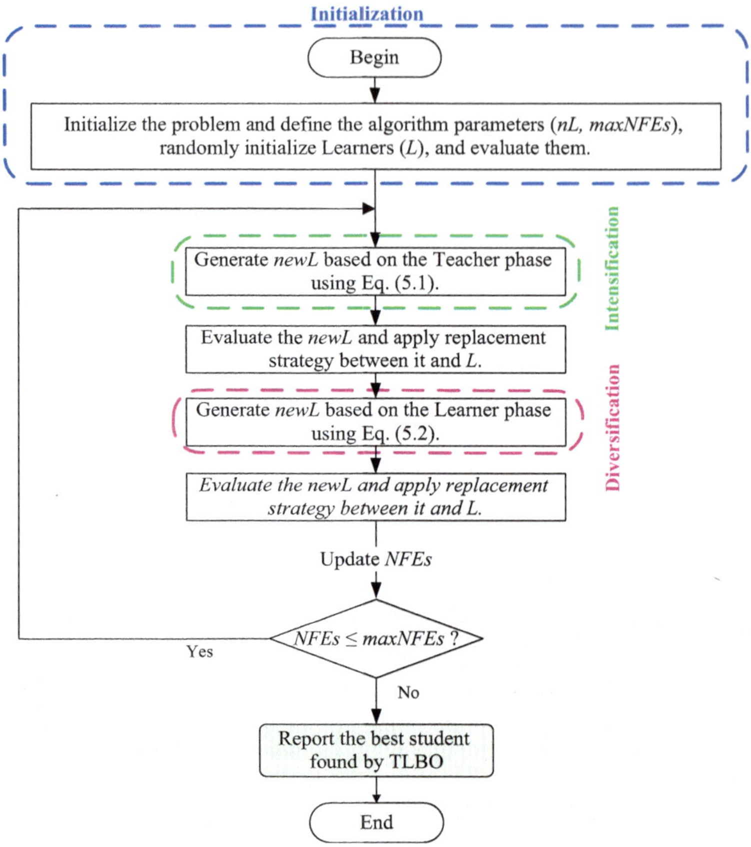

Fig. 5.1 Flowchart of the TLBO algorithm

5.3 MATLAB Code for the Teaching-Learning-Based Optimization Algorithm

The algorithm is coded here in a simple manner. According to the previous subsection, four functions are considered for coding the algorithm. Firstly, these functions are presented in the order of their recall in the algorithm, and at the end, the algorithm is coded. The required comments are presented. Note that the percent sign (%) represents a comment.

The first function is the objective function evaluation (*fobj*) which is presented in the second chapter.

The second function is the new students' generation function based on the teacher phase. This function is named as *Teacher*. The input arguments of this function are the current set of students (L) and the corresponding penalized objective function vector (*PFit*), and the output is the newly generated ones (*newL*). The teacher function is coded as follows:

```
% Get new Learners by the Teacher Phase

function newL=Teacher(L,PFit)

% Calculate the mean of position of students (column-wise)
for i=1:size(L,2)
    MeanL(1,i)=sum(L(:,i))/size(L,1);
end

% The best solution will act as a teacher for that iteration
[TPfit,k]=min(PFit);
T=L(k,:);% T: Teacher

for i=1:size(L,1)
    TF = randi([1 2]);
    stepsize(i,:)=T-TF.*MeanL;
end
newL=L+rand(size(L)).*stepsize;
```

The third function is the *Replacement* function in which the old students are compared and replaced with the newly educated ones in a simple greedy manner so that the student with smaller penalized objective function (*pfit*) is preferred. The *fobj* function is called in the nested form within this function. The input and output arguments for this function is the same as the *fobj* function but in the matrix form of all candidate solutions. *Replacement* function is coded in the following:

```
% Evaluating and updating Learners by comparing the old and new ones.

function [L,Fit,PFit]=Replacement(L,newL,Fit,PFit,Lb,Ub)

for i=1:size(L,1),
     [X,fit,pfit]=fobj(newL(i,:),Lb,Ub);
     if pfit<=PFit(i)
         L(i,:)=X;
         Fit(i)=fit;
         PFit(i)=pfit;
     end
end
```

The fourth function is the new students' generation function based on the learner phase. This function is named as *Learner*. The input and output arguments of this function are the same as the *Teacher* function. It should be noted that each student should interact with a randomly selected student except him or herself. This function is coded as follows:

```
% Get new Learners by the Learner phase.

function newL=Learner(L,PFit)

% ith learner interacts with randomly selected jth learner where i~=j.
rp=randperm(size(L,1));
for i=1:size(rp,2)
    while rp(i)==i
        rp(i)=round(1+(size(L,1)-1).*rand);
    end
end

for i=1:size(L,1)
    if PFit(i)<PFit(rp(i))
        stepsize(i,:)=L(i,:)-L(rp(i),:);
    else
        stepsize(i,:)=L(rp(i),:)-L(i,:);
    end
end
newL=L+rand(size(L)).*stepsize;
```

The TLBO algorithm is coded in the following as the composition of three parts, namely, initialization, algorithm cyclic body, and monitoring the results.

```
clc
clear

%% Initialization

% Define the properties of COP (tension/compression spring design problem).
nV=3; % Number of design variables.
Lb=[0.05 0.25 2]; % Lower bounds of design variables.
Ub=[2 1.3 15]; % Upper bounds of design variables.

% Define parameters of TLBO algorithm.
nL=20;% Number of Learners
maxNFEs=20000; % Maximum number of Objective Function Evaluations.


% Generate random initial solutions.
for i=1:nL
    L(i,:)=Lb+(Ub-Lb).*rand(1,nV); % Students or learners matrix or matrix
of the initial candidate solutions or the initial population.
end

% Evaluate initial population (L) calling the fobj function constructed in
the second chapter and form its corresponding vectors of objective function
(Fit) and penalized objective function (PFit). It should be noted that the
design vectors all are inside the search space.
for i=1:size(L,1)
    [X,fit,pfit]=fobj(L(i,:),Lb,Ub);
    L(i,:)=X;
    Fit(i)=fit;
    PFit(i)=pfit;
end

% Monitor the best candidate solution (bestL) and its corresponding
penalized objective function (minPFit) and objective function (minFit).
[minPFit,m]=min(PFit);
minFit=Fit(m);
bestL=L(m,:);

%% Algorithm Body

NFEs=0; % Current number of Objective Function Evaluations used by the
algorithm until yet.
NITs=0; % Number of algorithm iterations.

while NFEs<maxNFEs
    NITs=NITs+1; % Update the number of algorithm iterations.

    % Generate new Learners based on the Teacher phase
    newL=Teacher(L,PFit);

    % Replace the old Learners with the newly generated ones if the newest
ones are better. Notice that the fobj function is called in the following
replacement function in nested form. Hence the newly generated Learners
will be corrected and evaluated.
    [L,Fit,PFit]=Replacement(L,newL,Fit,PFit,Lb,Ub);

    % Update the number of Objective Function Evaluations used by the
algorithm until yet.
    NFEs=NFEs+nL;
```

```
    % Generate new Learners based on the Learner phase
    newL=Learner(L,PFit);

    % Replace the old Learners with the newly generated ones if the newest
ones are better. Notice that the fobj function is called in the following
replacement function in nested form. Hence the newly generated Learners
will be corrected and evaluated.
    [L,Fit,PFit]=Replacement(L,newL,Fit,PFit,Lb,Ub);

    % Update the number of Objective Function Evaluations used by the
algorithm until yet.
    NFEs=NFEs+nL;

    % Monitor the best candidate solution (bestL) and its corresponding
penalized objective function (minPFit) and objective function (minFit).
    [minPFit,m]=min(PFit);
    minFit=Fit(m);
    bestL=L(m,:);

    % Display desired information of the iteration.
    disp(['NITs= ' num2str(NITs) '; minFit = ' num2str(minFit) '; minPFit
= ' num2str(minPFit)]);

    % Save the required results for post processing and visualization of
algorithm performance.
    output1(NITs,:)=[minFit,minPFit,NFEs];
    output2(NITs,:)=[min(PFit),max(PFit),mean(PFit)];
    output3(NITs,:)=[bestL,NFEs];
end

%% Monitoring the results.
figure;
plot((1:1:NITs),output2(:,1),'g',(1:1:NITs),output2(:,2),'r--
',(1:1:NITs),output2(:,3),'b-.')
legend('min','max','mean');
xlabel('NITs');
ylabel('pfit');
axis([1 NITs 0.95*min(PFit) 1.2*min(PFit)])
```

5.4 Experimental Evaluation

As it is mentioned in the second chapter, the aim of experimental evaluation is to show the effects of algorithm parameters on its performance and study the exploration and exploitation capabilities of the algorithm. TLBO has two parameters: maximum number of objective function evaluations (*maxNFEs*) as the stopping criterion of the algorithm and the number of learners (*nL*). These two parameters are also used in any other algorithms. In this regard TLBO can be called as a nonparametric metaheuristic. Considering a large enough value for *maxNFEs* to ensure the convergence of an algorithm is essential. This parameter almost is problem dependent and should be considered larger for complicated and larger problems. It is considered equal to 20,000 for TLBO and other algorithms. However, to show the influence of population size on the algorithm performance, Fig. 5.2 shows the algorithm convergence histories of a single run of the algorithm with different values of *nL* from 5 to 200. The runs are performed from a similar random

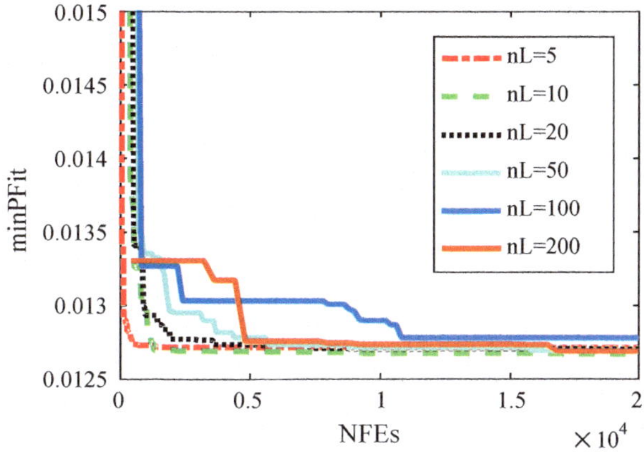

Fig. 5.2 Convergence histories of the algorithm for different values of *nL*

initial population. For further clarity the upper bound of Y axis is limited. It is clear that at least ten numbers of learners are needed to have an acceptable performance of the algorithm. Small number of learners results in much intensification. Although the algorithm has not been trapped in local optima for the small number of population, it can be problematic for most complex problems. The value of 50 results in a better performance in the both aspects of accuracy and convergence speed, and population size more than 100 takes the algorithm toward more diversification and results in slow convergence speed and less accuracy.

Reference

1. Rao RV, Savsani VJ, Vakharia DP (2011) Teaching–learning-based optimization: a novel method for constrained mechanical design optimization problems. Comput Aided Des 43 (3):303–315

Chapter 6
Imperialist Competitive Algorithm

6.1 Introduction

This chapter presents imperialist competitive algorithm (ICA) proposed by Atashpaz-Gargari and his co-workers [1, 2] which is a socio-politically motivated optimization algorithm. ICA was firstly used by Kaveh and Talatahari [3] for steel structural optimization. Like other population-based metaheuristics, ICA starts with a set of random initial candidate solutions, as the initial countries. A specific number of best countries is considered as the emperors which take a number of remaining countries as their colonies based on competency. Therefore, the countries are categorized into emperors and colonies and collectively form empires or imperialist states. ICA has two main mechanisms who are the basis of the ICA: improving the colonies of each empire by intrinsic learning of colonies from their emperor and imperialistic competitions among empires. First one results in powering empires themselves. In this way each colony has opportunity to take the role of emperor of that empire. During imperialistic competitions among empires, weakest empires lose their weakest colonies, and powerful empires take possession of them until the weakest empire collapses. The power of each imperialist or empire not only depends on the quality or position of its emperor and colonies but also depends on the number of its colony. Colonies improvements and imperialistic competitions direct the search process toward the powerful imperialists or the optimum points. These are repeated in the cyclic body of the algorithm in succession to satisfy a stopping criteria with the aim of collapsing all the empires except the most powerful one which will have all the countries under its control.

A. Kaveh, T. Bakhshpoori, *Metaheuristics: Outlines, MATLAB Codes and Examples*, https://doi.org/10.1007/978-3-030-04067-3_6

6.2 Formulation and Framework of the Imperialist Competitive Algorithm

ICA was inspired by the historical phenomenon of imperialism and colonialism. Looking at the colonial competition in human society, one can find an analogy for a population-based metaheuristic. A set of countries can be considered as the algorithm population, and their value of cost shows the competency to have imperialists. Some of the best countries can be considered as emperors which take a number of remaining countries as colonies based on competency. Therefore, the countries or individuals are categorized into emperors and colonies and collectively form empires or imperialist states. Each empire or imperialist composed of an emperor and a number of colonies in accordance with its competency. Each empire aims to enhance its power by two mechanisms: firstly, by improving its colonies and secondarily by taking the possession of other colonies from other empires. These two mechanisms are iterated in the cyclic body of the algorithm with the aim of converging to an ideal world in which the weakest empires are collapsed except the most powerful one. In this case all the countries have the same position and cost and are under control of one emperor with the same position and cost. This position and cost is the optimum solution found by the algorithm.

ICA is developed in four steps [1, 3]: forming the initial empires, colonies movement, emperor updating, and imperialistic competition. Except the first step, other three ones form the cyclic body of the algorithm. Like other population-based metaheuristics, ICA starts with a set of random initial solutions or countries. Consider nC number of countries as the population matrix or the countries matrix (C). Evaluating the C matrix its corresponding objective function (Fit) and penalized objective function ($PFit$) vectors can be formed.

First Step: Forming the Initial Empires

In this step a specific number (nE) of the best countries (in constrained optimization terminology, countries with the lower penalized objective functions) will be selected to be the emperors. The matrix of emperors and its corresponding objective function and penalized objective function vectors are called eC, $eFit$, and $ePFit$, respectively. It should be noted that these are of the size of $nE \times nV$, $1 \times nE$, and $1 \times nE$, respectively. The remaining countries will form the colonies of these imperialists. In order to proportionally divide the colonies among the emperors, a normalized cost (NC) for an emperor is defined as:

$$NC(i) = ePFit_i - \max(ePFit), \quad i = 1, \ldots, nE \tag{6.1}$$

where max function returns the maximum value of penalized objective function of emperors as the worse one. Computing the normalized cost of emperors, normalized power (NP) of each emperor is defined as:

$$NP(i) = \left| \frac{NC(i)}{\mathrm{sum}(NC)} \right|, \quad i = 1, \ldots, nE \tag{6.2}$$

where sum function calculates the summation of normalized cost of all emperors. In fact, *NP* is the portion of colonies that should be possessed by the *i*th emperor. The initial number of colonies (*nCOL*) of each empire is as follows:

$$nCOL(i) = \mathrm{round}(NP(i) \times (nC - nE)), \quad i = 1, \ldots, nE \tag{6.3}$$

where round is the rounding function and $nC - nE$ is the number of all colonies. To divide the colonies, for each emperor *nCOL(i)* colonies are randomly selected and assigned to it to form the *i*th empire. It was mentioned in the second chapter that we will try to code algorithms in a simple manner and avoid to use even cell arrays. However, in the ICA algorithm, inevitably empires and their corresponding objective function and penalized objective function are defined as cell arrays and called *EC*, *EFit*, and *EPFit*, respectively.

Equations (6.1–6.3) are formulated in the original ICA. However, there are some problems with this formulation. Firstly, based on the Eq. (6.1), the worse emperor will be assigned by the normalized cost of zero and as a result form an empire without any colony. Secondly, based on the Eq. (6.3) in most cases, cumulative summation of *nCOL* is less or more than the number of all colonies ($nC - nE$) because of using the rounding function. To address the first problem firstly, Eq. (6.1) is considered as follows:

$$NC(i) = ePFit_i - \max(PFit), \quad i = 1, \ldots, nE \tag{6.4}$$

and for the latter the fitness proportionate selection or roulette wheel selection based on the normalized power vector is used instead of the Eq. (6.3). For example, the results for *nCOL* obtained by these two formulations considering different numbers of *nC* and *nE* are tabulated in the following Table 6.1. It should be noted that we had to use the Eq. (6.4) also in the ICA original formulation. As it is clear the new formulation solves both the problems. The point worth to mention is that original formulation of ICA results in values for *nCOL* which are descending.

Table 6.1 Calculating the number of colonies for each empire (*nCOL*)

Case	Formulation	*nCOL*					Sum (*nCOL*)
$nC = 50$	ICA	17	16	13	–	–	46
$nE = 3$	Corrected	16	16	15	–	–	47
$nC = 50$	ICA	17	12	9	9	–	47
$nE = 4$	Corrected	17	14	6	9		46
$nC = 50$	ICA	10	9	9	8	8	44
$nE = 5$	Corrected	11	6	8	12	8	45

Second Step: Colonies Movement

In the ICA, emperors start to improve their colonies. This is modeled by moving the colonies of each empire toward its emperor. In fact, the assimilation policy pursued by some of former imperialist states are modeled by moving all the colonies toward the imperialist. It should be noted that any colony can take the place of its emperor if it visits a better position in the search space. The movement of *j*th colony from the *i*th empire ($EC\{i\}(j)$) takes place by a uniform random direction toward their corresponding emperor ($eC(i)$) and a uniform random deviation which can be formulated as:

$$\begin{gathered} \text{direction} = \beta \times U(0, (EC\{i\}(j) - eC(i))) \\ i = 1, \ldots, nE; \text{ and } j = 1, \ldots, nCOL(i) \end{gathered} \tag{6.5}$$

$$\begin{gathered} \text{deviation} = U(-\gamma, +\gamma) \\ i = 1, \ldots, nE; \text{ and } j = 1, \ldots, nCOL(i) \end{gathered} \tag{6.6}$$

$$newEC = EC + \text{direction} * \text{deviation} \tag{6.7}$$

in which $newEC$ is the new position of the colonies and $U\,(a,\,b)$ is the uniformly random value generation function distributed between a and b. β is a parameter with a value greater than one. γ is a parameter that adjusts the deviation from the original direction. In most of the implementations, a value of about 2 for β and about $\pi/4$ (Rad) for γ result in a good convergence of the colonies to the emperor. It should be noted that many improvements developed for this step of ICA which are not the subject of this book.

Third Step: Emperor Updating

If the new position of the colony is better than that of its relevant imperialist (considering the penalized objective function), the imperialist and the colony change their positions, and the new location with a lower cost becomes the imperialist. Then in the next iteration, the other colonies move toward this new position.

Forth Step: Imperialistic Competition

Imperialistic competition is another strategy utilized in the ICA methodology. All empires try to take the possession of colonies of other empires and control them. The imperialistic competition gradually reduces the power of weaker empires and increases the power of more powerful ones. The imperialistic competition is modeled by just picking some (usually one) of the weakest colonies of the weakest empires and making a competition among all empires to possess these (this) colonies. In this competition based on their total power, each of empires will have a possibility of taking possession of the mentioned colonies. Total power of an empire is mainly affected by the power of imperialist country or the emperor. But the power of the colonies of an empire has an effect, though negligible, on the total power of that empire. This fact is modeled by defining the total cost as:

$$TC(i) = ePFit(i) + \xi \cdot \text{mean}(EPFit\{i\}), i = 1, \ldots, nE \tag{6.8}$$

where *TC*(i) is the total cost of the ith empire, mean is a function that returns the mean or average value of the penalized objective function of the colonies of the ith empire, and ξ is a positive number which is considered to be less than 1. A small value for ξ causes the total power of the empire to be determined by just the imperialist, and increasingly it will add to the role of the colonies in determining the total power of the corresponding empire. The value of 0.1 for ξ is found to be a suitable value in most of the implementations. Similar to Eq. (6.2), the normalized total cost is defined as:

$$NTC(i) = TC(i) + \max(TC), i = 1, \ldots, nE \tag{6.9}$$

where *NTC*(i) is the normalized total cost of the ith empire. Having the normalized total cost, the possession probability of each empire is evaluated by:

$$P(i) = \frac{NTC(i)}{\text{sum}(NTC)}, i = 1, \ldots, nE \tag{6.10}$$

where sum function calculates the summation of normalized total cost of all empires. With the same size of possession vector, a uniformly random vector from the [0, 1] interval generated and subtracted with it, and a possession index vector (*D*) formed as:

$$D(i) = P(i) - rand(0, 1), i = 1, \ldots, nE \tag{6.11}$$

The empire whose relevant index is maximum in the *D* will hand the weakest colony from the weakest empire. It should be noted that in all three steps, colonies movement, emperor updating, and imperialistic competition, the power of empires which is characterized by total cost is variable. However, the total cost is needed only in the imperialistic competition step, and we had to calculate the *TC* vector only in this step.

The point worth to mention is that when an empire loses all of its colonies, it is assumed to be collapsed. In this regard the emperor itself will be also possessed by the empire with the maximum *D* value. At the end it should be noted that when a colony is assigned to the winner empire, it is certainly worse than the emperor of that empire. Then the emperor updating is not needed to be checked. For the collapsed empire case, it is not also needed. Also consider that the imperialist competition will happen when there is more than one empire.

The pseudo code of algorithm is provided as follows, and the flowchart of ICA is illustrated in Fig. 6.1.

The pseudo code of the ICA algorithm for solving COPs:

Define the algorithm parameters: *nC*, *nE*, β, γ, ξ, and *maxNFEs*.
Generate random initial solutions or countries (*C*).

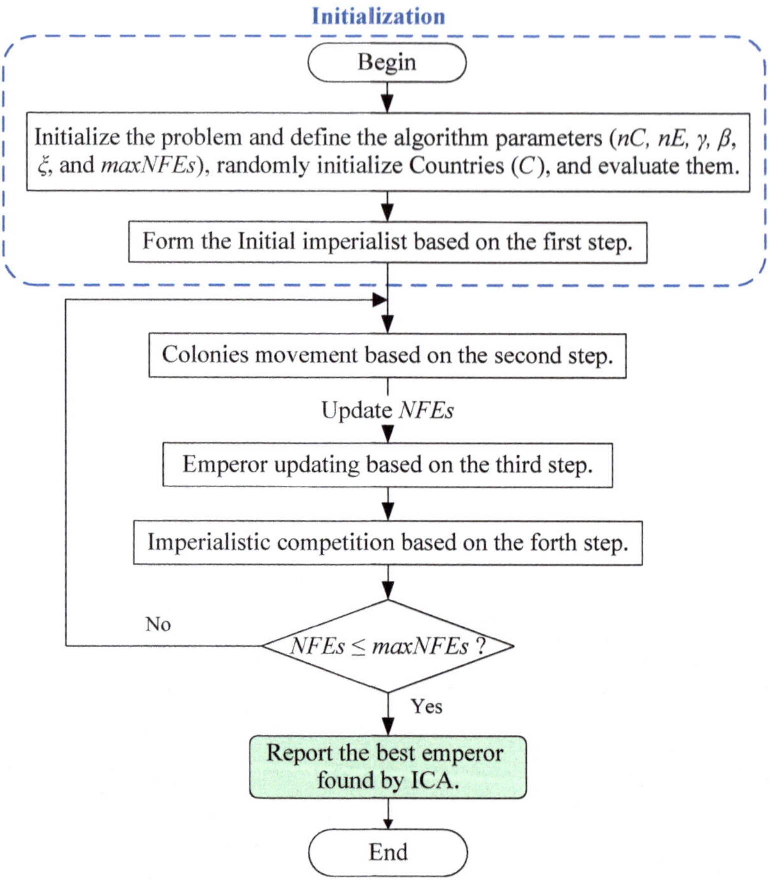

Fig. 6.1 Flowchart of the ICA algorithm

Evaluate initial population (*C*), and form its corresponding vectors of objective function (*Fit*) and penalized objective function (*PFit*).
Form the initial imperialist.
While *NFEs*<*maxNFEs*

Update the number of algorithm iterations (*NITs*).
Move colonies toward the emperor within each empire and evaluate them.
Update *NFEs*.
Change the new colony with its emperor, if the new position of the colony is better than that of its relevant imperialist.
Pick the weakest colony of weakest empire, and give it to the most powerful empire. Check it if the weakest colony is collapsed, and then give its emperor to the most powerful empire.
Monitor the best country which in fact is the best emperor.

end While

According to this flowchart, it is obvious that one cannot explicitly make a distinguish between intensification and diversification because the ICA involves these two features together in the colonies movement step using direction and diversification components.

6.3 MATLAB Code for the Imperialist Competitive Algorithm (ICA)

It should be noted that to code the algorithm in a simple manner so that the reader can trace the code line by line and even easily realize all the items defined or generated in MATLAB environment, using ready functions and structure arrays are avoided. According to the previous section, five functions are considered for coding the algorithm. First, these functions are presented with the order of their recall in the algorithm and at the end, the algorithm is coded. The required comments are presented. Note that the percent sign (%) represents a comment.

The first function is the objective function evaluation (*fobj*) which was presented in Chap. 2.

The second function named as the Initial Imperialist Competition (*ICC*) function to form the initial imperialist based on the first step. The input arguments are the initial set of countries matrix (*C*), its correspondence objective function (*Fit*) and penalized objective function (*PFit*) vectors, and the number of empires (*nE*). The output arguments are the matrix of emperors (*eC*), its correspondence objective function (*eFit*) and penalized objective function (*ePFit*) vectors, the cell of empires (*EC*) and its correspondence objective function (*EFit*) and penalized objective function (*EPFit*) cells, and a vector to show the number of colonies of empires (*nCOL*). It should be noted that normalized cost of emperors and other vectors are needed in this step to determine the number of colonies for each empire and should be calculated. The ICC function is coded as follows.

```
% Forming the initial imperialist based on the ICC function

function [eC,eFit,ePFit,EC,EFit,EPFit,nCOL]=IIC(C,Fit,PFit,nE)
ncol=size(C,1)-nE;% number of colonies.

% Form the matrix of emperors (eC) and its corresponding objective (eFit)
and penalized objective function (ePFit) vectors.
[~,order1]=sort(PFit);
eC(1:nE,:)=C(order1(1:nE),:);
eFit(1:nE)=Fit(order1(1:nE));
ePFit(1:nE)=PFit(order1(1:nE));

% Compute the normalized cost (NC) and then normalized power (NP) of each
emperor and at the end its initial number of colonies (nCOL) using the
fitness proportionate selection or roulette wheel selection.
NC=ePFit-max(PFit);
NP=abs(NC/sum(NC));
nCOL=zeros(1,nE);
for i=1:ncol
    j=find(rand<=cumsum(NP),1,'first');
    nCOL(j)=nCOL(j)+1;
end

% For each emperor, nCOL number of colonies are randomly selected from the
initial population and assigned to it to form its empire (EC. Accordingly
form objective function vector (EFit) and penalized objective function
vector(EPFit) for each empire.
order2=randperm(ncol);
k=0;
for i=1:nE
    for j=1:nCOL(i)
        k=k+1;
        EC{i}(j,:)=C(nE+order2(k),:);
        EFit{i}(j)=Fit(nE+order2(k));
        EPFit{i}(j)=PFit(nE+order2(k));
    end
end
```

The third function is considered to move colonies toward their emperors based on the third step and named as *Colonies_Movement*. The input arguments of this function are all outputs of the ICC function, current number of function evaluations (*NFEs*), parameters used in the movement strategy (γ and β), and the lower and upper bound of design variables (*Lb* and *Ub*). The output arguments are the same as the ICC function except the last argument that has come *NFES* instead *nCOL*, because the *nCOL* is unchanged in the movement step. The *Colonies_Movement* function is coded as follows.

```
% moving colonies toward their emperors based on the Colonies_Movement
function

function
[eC,eFit,ePFit,EC,EFit,EPFit,NFEs]=Colonies_Movement(eC,eFit,ePFit,EC,EFit,
EPFit,nCOL,NFEs,beta,gamma,Lb,Ub)

nE=size(eC,1);

for i=1:nE
    for j=1:nCOL(i)
        d=eC(i,:)-EC{i}(j,:);
        direction=beta.*d.*rand(size(EC{i}(j,:)));
        deviation=-gamma+2*gamma.*rand(size(EC{i}(j,:)));
        EC{i}(j,:)= EC{i}(j,:)+direction.*deviation;
        % Evaluate moved colony.
        [EC{i}(j,:),EFit{i}(j),EPFit{i}(j)]=fobj(EC{i}(j,:),Lb,Ub);
        % Update the number of function evaluations.
        NFEs=NFEs+1;
    end
end
```

The forth function changes the best new colony or each empire with its emperor (considering the penalized objective function) based on the third step. This function is named as *Emperor_Updating* function. Input and output arguments of this function are the output arguments of the IIC function except the last argument which is unchanged in this step. The *Emperor_Updating* function is coded as follows.

```
% Changing the best new colony with its emperor (considering the PFit)
based on the Emperor_Updating function

function
[eC,eFit,ePFit,EC,EFit,EPFit]=Emperor_Updating(eC,eFit,ePFit,EC,EFit,EPFit)

nE=size(eC,1);

for i=1:nE
    [new_ePFit,index]=min(EPFit{i});
    new_eC=EC{i}(index,:);
    new_eFit=EFit{i}(index);
    if new_ePFit<=ePFit(i)
        EC{i}(index,:)=eC(i,:);
        EFit{i}(index)=eFit(i);
        EPFit{i}(index)=ePFit(i);
        eC(i,:)=new_eC;
        eFit(i)=new_eFit;
        ePFit(i)=new_ePFit;
    end
end
```

The last function is the *Imperialistic_Competition* function by which the possession of the weakest colony from the weakest emperor will be taken by the most powerful empire. It should be noted that the total cost of empires should be calculated firstly in this step to determine the most powerful empire (indexed here by *a*) to pick the weakest colony (indexed here by *c*) from the weakest empire (indexed here by *b*). It should be noted also that the collapse of weakest empires should be checked and applied if it would happen. The input arguments are the same as the output arguments of IIC function plus another one: the ξ parameter to control the contribution of colonies and emperor of each empire in determining its power. Output arguments are same as the IIC function. The *Imperialistic_Competition* function is coded as follows.

```
% Taking the possession of the weakest colony from the weakest emperor
based on the Imperialistic_Competition function.

function
[eC,eFit,ePFit,EC,EFit,EPFit,nCOL]=Imperialistic_Competition(eC,eFit,ePFit,
EC,EFit,EPFit,nCOL,kesi)

nE=size(eC,1);

% Calculate the total cost of empires (ETC).
for i=1:nE
    ETC(i)=ePFit(i)+kesi*mean(EPFit{i});
end

% Compute the normalized total cost (NTC) and then possession probability
(PP) of each emperor and at the end the competition index vector (D).
NTC=ETC-max(ETC);
PP=abs(NTC/sum(NTC));
D=PP-rand(size(PP));

% Determine the index of winner emperor (a) which is the one with maximum
value of the competition index (D)
[~,a]=max(D);

% Determine the index of weakest empire (b) and index of its weakest colony
(c). It will be possessed by the winner emperor. Notice that if an emperor
loses all of its colonies it will be assumed as a collapsed empire and
itself will be assigned also to the winner empire.
[~,b]=max(ePFit);

if nCOL(b)>0
    [~,c]=max(EPFit{b});
    EC{a}(end+1,:)=EC{b}(c,:);
    EFit{a}(end+1)=EFit{b}(c);
    EPFit{a}(end+1)=EPFit{b}(c);
    nCOL(a)=nCOL(a)+1;
    EC{b}(c,:)=[];
    EFit{b}(c)=[];
    EPFit{b}(c)=[];
    nCOL(b)=nCOL(b)-1;
end

if nCOL(b)==0
    EC{a}(end+1,:)=eC(b,:);
    EFit{a}(end+1)=eFit(b);
    EPFit{a}(end+1)=ePFit(b);
    nCOL(a)=nCOL(a)+1;
    EC(b)=[]; EFit(b)=[]; EPFit(b)=[];
    eC(b,:)=[]; eFit(b)=[]; ePFit(b)=[];
    nCOL(b)=[];
    ETC(b)=[];
end
```

The ICA algorithm is coded in the following composed of three parts: initialization, algorithm cyclic body, and monitoring the results.

```
clc
clear

%% Initialization

% Define the properties of COP (tension/compression spring design problem).
nV=3; % Number of design variables.
Lb=[0.05 0.25 2]; % Lower bounds of design variables.
Ub=[2 1.3 15]; % Upper bounds of design variables.

% Define parameters of ICA algorithm.
nC=20; % Number of Countries
nE=3; % Number of Empires.
beta=2; % Parameter to control the length of movement of colonies toward
their emperors.
gamma=pi/4; % Parameter to adjust deviation of the direction of movement of
colonies toward their emperors.
kesi=0.1; % Parameter to determine the total cost of an empire.
maxNFEs=20000; % Maximum number of Objective Function Evaluations.

% Generate random initial solutions.
for i=1:nC
    C(i,:)=Lb+(Ub-Lb).*rand(1,nV); % Countries matrix or matrix of the
initial candidate solutions or the initial population.
end

% Evaluate initial population (C) calling the fobj function constructed in
the second chapter and form its corresponding vectors of objective function
(Fit) and penalized objective function (PFit). It should be noted that the
design vectors all are inside the search space.
for i=1:nC
    [X,fit,pfit]=fobj(C(i,:),Lb,Ub);
    C(i,:)=X;
    Fit(i)=fit;
    PFit(i)=pfit;
end

% Form the initial imperialist using the Initial Imperialist Competition
(IIC) function.
[eC,eFit,ePFit,EC,EFit,EPFit,nCOL]=IIC(C,Fit,PFit,nE);


% Monitor the best candidate solution (bestC) and its corresponding
objective function (minFit) and penalized objective function (minPFit). It
should be noted that the bestC is the best emperor.
[minPFit,m]=min(ePFit);
minFit=eFit(m);
bestC=eC(m,:);

%% Algorithm Body

NFEs=0; % Current number of Objective Function Evaluations used by the
algorithm until yet.
NITs=0; % Number of algorithm iterations

while NFEs<maxNFEs
    NITs=NITs+1; % Update the number of algorithm iterations.

    % Colonies would be moved toward the emperor based on the Colonies
Movement function.
[eC,eFit,ePFit,EC,EFit,EPFit,NFEs]=Colonies_Movement(eC,eFit,ePFit,EC,EFit,
EPFit,nCOL,NFEs,beta,gamma,Lb,Ub);
```

```
    % Emperor updating.
[eC,eFit,ePFit,EC,EFit,EPFit]=Emperor_Updating(eC,eFit,ePFit,EC,EFit,EPFit)
;

    if size(eC,1)~=1
        % Imperialistic competition.
[eC,eFit,ePFit,EC,EFit,EPFit,nCOL]=Imperialistic_Competition(eC,eFit,ePFit,
EC,EFit,EPFit,nCOL,kesi);
    end


    % Monitor the best candidate solution (bestC) and its corresponding
penalized objective function (minPFit) and objective function (minFit). It
should be noted that the bestC is the best emperor.
    [minPFit,m]=min(ePFit);
    minFit=eFit(m);
    bestC=eC(m,:);

    % Display desired information of the iteration.
    disp(['NITs= ' num2str(NITs) '; minFit = ' num2str(minFit) '; minPFit
= ' num2str(minPFit)]);

    % Save the required results for post processing and visualization of
algorithm performance.
    output1(NITs,:)=[minFit,minPFit,NFEs];

output2(NITs,:)=[minFit,max([max(ePFit),max(cell2mat(EPFit))]),mean([mean(e
PFit),mean(cell2mat(EPFit))])];
    output3(NITs,:)=[bestC,NFEs];
end

%% Monitoring the results
figure;
plot((1:1:NITs),output2(:,1),'g',(1:1:NITs),output2(:,2),'r--
',(1:1:NITs),output2(:,3),'b-.')
legend('min','max','mean');
xlabel('NITs');
ylabel('PFit');
```

6.4 Experimental Evaluation

The ICA has six parameters: number of countries (*nC*), number of empires (*nE*), parameter to control the length of movement of colonies toward their emperors (γ), parameter to adjust deviation of the direction of movement of colonies toward their emperors (β), parameter to control the contribution of emperor and its colonies in calculating the power of an empire (ξ), and maximum number of objective function evaluations as the stopping criteria (*maxNFEs*). Considering a large enough value for *maxNFEs* to ensure convergence of the algorithms is essential. This parameter is almost problem dependent and should be considered larger for complicated and larger problems. It is considered equal to 20,000 for ICA like previous chapters.

γ and β parameters are considered as 2 and $\pi/4$, respectively. These values are recommended in the available literature of different applications of ICA and result in acceptable performance of the algorithm. Figure 6.2 depicts the algorithm performance for different values of γ and β parameters considering *nC*, *nE*, and ξ equal to 20, 3, and 0.1, respectively. A single run has been done for each case from a fixed

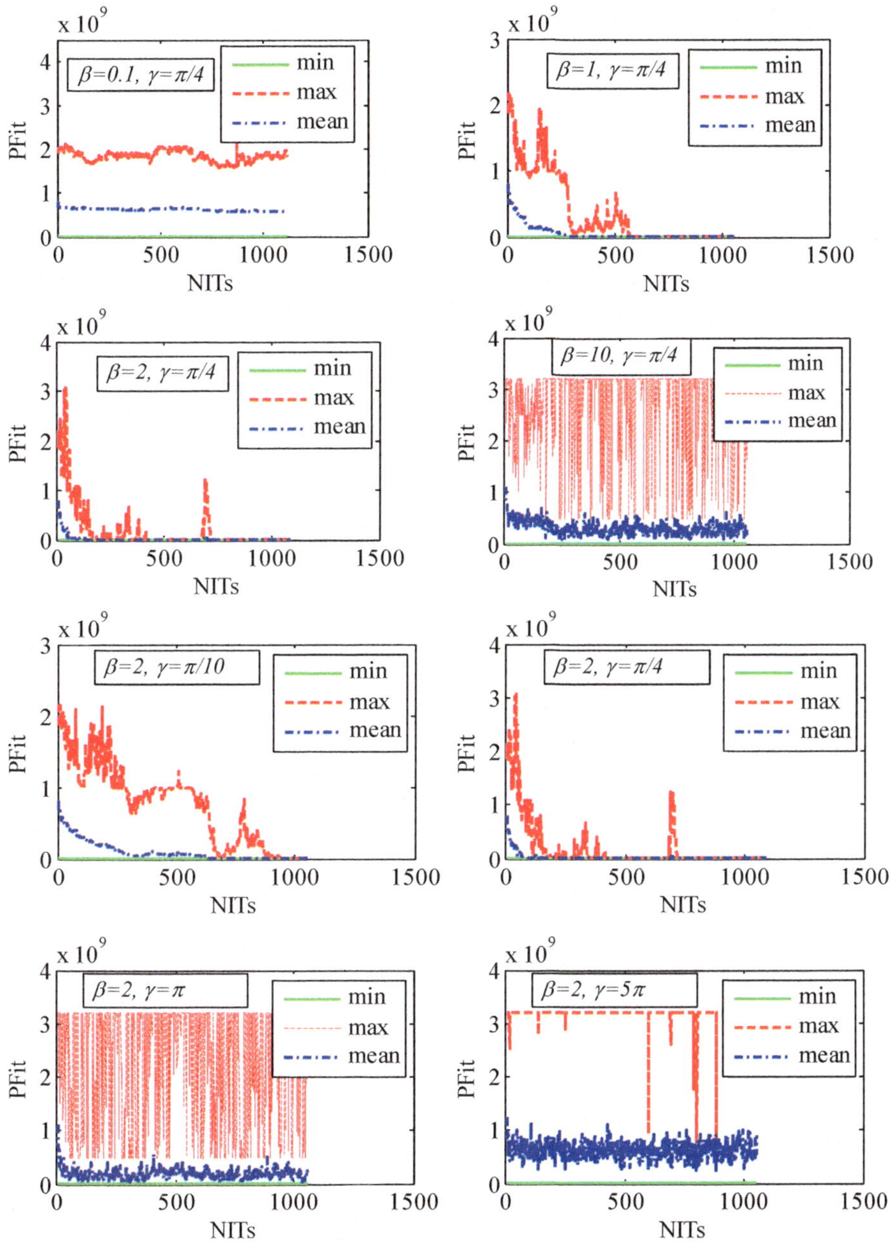

Fig. 6.2 Convergence histories of the minimum, maximum, and mean of the algorithm population considering different values for γ and β parameters

randomly initial population, and convergence histories of the best, worst, and mean values of the penalized objective function of all countries are monitored. It is clear that $\beta > 1$ makes the colonies to move closer to the emperor state from both sides. $\beta \gg 1$ gradually results in a divergence of colonies from the emperor state, while a very close value to 1 for β reduces the search ability of the algorithm. It is also apparent that considering γ parameter about $\pi/4$ results in a good convergence of the countries to the global minimum. It should be noted that in all cases the min curve can converge to the global optimum. However, just in the case of $\beta = 2$ and $\gamma = \pi/4$, ICA can direct all of its countries to the global minimum. It should be noted that the number of algorithm iteration is more than 1000 (*maxNFEs*/*nC*) since the emperors will not be moved and they do not need to be evaluated.

To study the effect of the ξ parameter on the algorithm performance, ICA is started from a fixed initial population considering nC, nE, γ, and β equal to 20, 3, $\pi/4$, and 2, respectively, with different values of ξ. Convergence histories for the best, worst, and the average of penalized objective function of the population are depicted in the Fig. 6.3. It is clear that considering values less than unity for this parameter is requisite to reach a better convergence and diversification. In the case of $\xi = 0$, algorithm shows the fastest convergence speed. However, it is needed more diversification for complicated problems. The point is worth to mention is that algorithm overall performance have been unchanged for larger values of ξ.

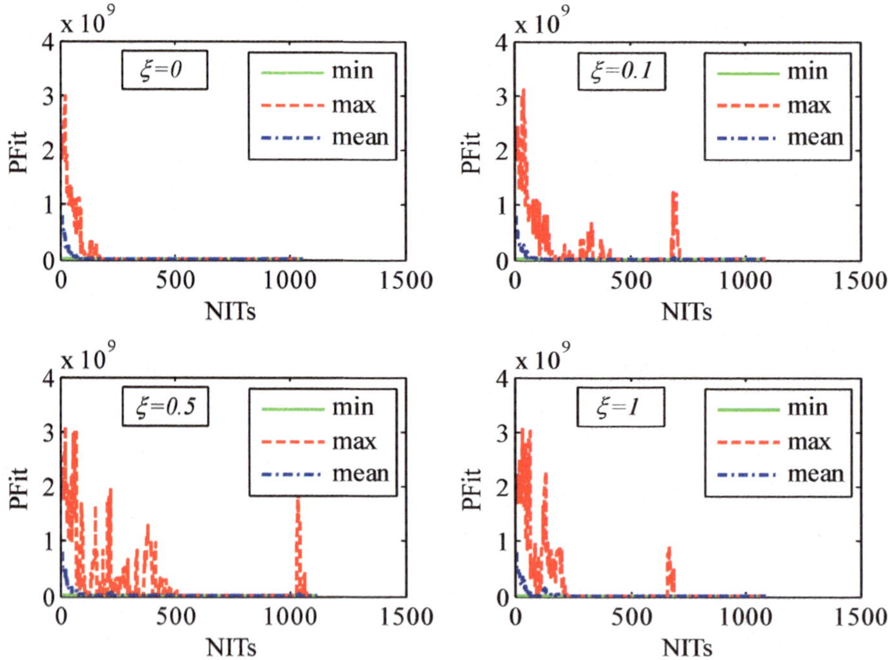

Fig. 6.3 Convergence histories of the minimum, maximum, and mean of the algorithm population considering different values for γ and β: (**a**) 20, (**b**) 50

The simulation results show that considering at least ten numbers of countries and three numbers of empires results in the better performance of the algorithm. Considering more numbers of countries results in more diversification of the algorithm and on the other hand needs more number of empires.

References

1. Atashpaz-Gargari E, Lucas C (2017) Imperialist competitive algorithm: an algorithm for optimization inspired by imperialistic competition. In: IEEE Congress on Evolutionary Computation, CEC 2007. IEEE
2. Atashpaz-Gargari E et al (2008) Colonial competitive algorithm: a novel approach for PID controller design in MIMO distillation column process. Int J Intell Comput Cybernet 1:337–355
3. Kaveh A, Talatahari S (2010) Optimum design of skeletal structures using imperialist competitive algorithm. Comput Struct 88:1220–1229

Chapter 7
Cuckoo Search Algorithm

7.1 Introduction

Cuckoo Search (CS) algorithm was developed by Yang and Deb [1, 2] as an efficient population-based metaheuristic inspired by the behavior of some cuckoo species in combination with the Lévy flight. It is also used in steel structural optimization problems by Kaveh and Bakhshpoori [3, 4]. Cuckoos are fascinating birds because of their special lifestyle and aggressive reproduction strategy. These species lay their eggs in the nests of other host birds with amazing abilities like selecting the recently spawned nests and removing existing eggs that increase the hatching probability of their eggs. The host takes care of the eggs presuming that the eggs are its own. However, some of host birds are able to combat with this parasites behavior of cuckoos and throw out the discovered alien eggs or build their new nests in new locations. Metaheuristics update the candidate solutions repetitively using the step sizes generated in the search space based on the inspired mechanisms or formulations. The randomization plays an important role in both exploration and exploitation in metaheuristic algorithms. Therefore, these step sizes are also combined with a series of consecutive random steps. In fact, the randomization part of the step sizes can vary according to a known distribution. A very special case can be achieved when the step length obeys the Lévy distribution and is called Lévy flight. The cuckoo breeding analogy combining with the Lévy flight behavior is the essence of the CS. Each solution represents a nest or an egg or a bird. Like other metaheuristics CS starts from randomly generated initial candidate solutions. These are the eggs of host birds. In the cyclic body of the algorithm, cuckoos aim to improve the quality of the solutions by generating new eggs with step sizes toward the best known solution in combination with the Lévy flight and intruded them to the nests of host birds. Moreover, in each iteration, after generating the new eggs by cuckoos, host birds have a chance to determine the alien eggs, abandon them, and generate new ones in the search space. Discovering the cuckoo's eggs and building new nests instead of

A. Kaveh, T. Bakhshpoori, *Metaheuristics: Outlines, MATLAB Codes and Examples*, https://doi.org/10.1007/978-3-030-04067-3_7

them can take place with different mechanisms. For example, it can take place for a fraction of eggs.

7.2 Formulation and Framework of the Cuckoo Search Algorithm

Looking at the parasitic lifestyle of cuckoo species and tackling of host birds against them, there is an analogy for developing a population-based metaheuristic inspiring by this biological system. All the nests or eggs whether they belong to the cuckoos or host birds represent the candidate solutions in the search space. Cuckoos and host birds try to breed their own generation.

CS starts with a set of randomly generated solutions. These are the nests or eggs of the host birds. The number of nests or eggs (*nN*) is the first parameter of the algorithm. *nN* numbers of nests form the *Nest* matrix. After evaluation of the initial population, the corresponding objective (*Fit*) and penalized objective function (*PFit*) vectors are also generated. In the cyclic body of the algorithm, two sequential search phases are performed by cuckoos and host birds. Firstly, cuckoos produce the eggs. In this phase eggs are produced by guiding the current solutions (*Nest*) toward the best known solution (*bestNest*) in combination with the Lévy flight. Then these new eggs are intruded to the nests of host birds based on the replacement strategy. After cuckoo breeding, it turns to the host birds. If a cuckoo's egg is very similar to a host's egg, then this cuckoo's egg is less likely to be discovered. In this phase host birds discover *pa* fraction of alien eggs and update them by adding them a random permutation-based step size. Based on the replacement strategy, host bird replaces the produced egg with the current one. For simplicity this can be done for all current eggs by considering a probability of updating matrix.

These two search phases are repeated in the cyclic body of the algorithm until reaching to the maximum number of objective function evaluations (*maxNFEs*) as the stopping criteria. After search process in each phase, replacement strategy will be performed to keep the old eggs or replace them with the newly generated ones. Different replacement strategies can be used in solving COPs. The simplest greedy strategy is used in this book consequently the solution with better quality or smallest *PFit* will be preferred to the old one. It should be noted that evaluation of new eggs firstly should be performed in replacement strategy; hence the number of objective function evaluations (*NFEs*) will be updated. In the following after introducing the Lévy flight, CS is formulated in two phases [1].

Lévy Flights as Random Walks

The randomization plays an important role in both exploration and exploitation in metaheuristic algorithms. The essence of such randomization is random walks. A random walk is a random process which consists of taking a series of consecutive random steps. Let S_N denote the sum of each consecutive random step X_i; then S_N forms a random walk:

$$S_N = \sum_{i=1}^{N} X_i = X_1 + X_2 + \ldots + X_N = \sum_{i=1}^{N-1} X_i + X_N = S_{N-1} + X_N \tag{7.1}$$

where X_i is a random step drawn from a random distribution, which means the next state will only depend on the current existing state and the motion or transition X_N from the existing state to the next state. If each step is carried out in the n-dimensional space, the random walk becomes in higher dimensions. In addition, there is no reason why each step length should be fixed. In fact, the step size can also vary according to a known distribution. For example, if the step length obeys the Gaussian distribution, the random walk becomes the Brownian motion. A very special case is when the step length obeys the Levy distribution, such a random walk is called a Lévy flight or Lévy walk.

From the implementation point of view, the generation of random numbers with Lévy flights consists of two steps: the choice of a random direction and the generation of steps which obey the chosen Lévy distribution, while the generation of steps is quite tricky. There are a few ways for achieving this, but one of the most efficient and yet straightforward ways is to use the so-called Mantegna algorithm. In Mantegna's algorithm, the step length S can be calculated by

$$S = \frac{u}{|v|^{1/\beta}} \tag{7.2}$$

where β is a parameter between [1, 2] interval and recommended in the literature to be 1.5; u and v are drawn from normal distribution. That is

$$u \sim N\left(0, \sigma_u^2\right), v \sim N\left(0, \sigma_v^2\right) \tag{7.3}$$

where

$$\sigma_u = \left\{ \frac{\Gamma(1+\beta) \times \sin(\pi\beta/2)}{\Gamma((1+\beta)/2) \times \beta \times 2^{(\beta-1)/2}} \right\}, \sigma_v = 1 \tag{7.4}$$

Studies show that the Lévy flights can maximize the efficiency of the resource searches in uncertain environments. In fact, Lévy flights have been observed among foraging patterns of albatrosses, fruit flies, and spider monkeys.

Cuckoo Breeding

In this step, all the nests or eggs except the best one (*bestNest*) are replaced based on their quality by new cuckoo eggs (*newNest*) produced by guiding the current solutions (*Nest*) toward the *bestNest* in combination with the Lévy flight as:

$$\begin{aligned} \text{stepsize} &= rand_{(i)(j)} \times \alpha \times S \times (Nest - bestNest) \\ newNest &= Nest + \text{stepsize} \end{aligned} \tag{7.5}$$

where α is the step size parameter and should be considered more than zero and should be related to the scales of the problem; $rand_{(i)(j)}$ is a random number chosen from the continuous uniform distribution on the $[-1, 1]$ interval, and S is a random walk based on the Lévy flights. This phase guarantees the elitism and intensification ability of the algorithm. The *bestNest* is kept unchanged and other solutions updated toward it.

Alien Eggs Discovery by the Host Birds

The alien eggs discovery is performed for each component of each solution in terms of the discovering probability matrix (P) such as:

$$P_{(i)(j)} = \begin{cases} 1 & \text{if } rand < pa \\ 0 & \text{if } rand \geq pa \end{cases} \tag{7.6}$$

where rand is a random number in [0, 1] interval and pa is the discovering probability. It should be noted that the P matrix has the same size as the *Nest* matrix. Existing eggs are replaced considering their quality by the newly generated ones from their current positions through random walks with a random permutation-based step size such as:

$$\begin{aligned} stepsize &= rand_{(i)(j)} \times (Nest[permute1(i)(j) - Nest[permute2(i)(j)]) \\ newNest &= Nest + stepsize \times P \end{aligned} \tag{7.7}$$

where *randperm*1 and *randperm*2 are random permutation functions used for different rows permutation applied on nest matrix and P is the discovery probability matrix. This phase guarantees the diversification ability of the algorithm.

The pseudo code of CS is provided as follows and the flowchart is illustrated in Fig. 7.1.

The pseudo code of the CS algorithm for solving COPs:

Define the algorithm parameters: nN, pa, α, and $maxNFEs$.
Generate random initial solutions or nests (*Nest*).
Evaluate initial population (*Nest*), form its corresponding vectors of objective function (*Fit*) and penalized objective function (*PFit*), and determine the best nest or egg found until yet (*bestNest*).
While $NFEs<maxNFEs$

Update the number of algorithm iterations (*NITs*).
Generate new eggs or nests (*newNest*) by cuckoos.
Evaluate the *newNest* and apply replacement strategy between old and new eggs.
Update *NFEs*.
Discovering the alien eggs by the host birds and generation of the new eggs or nests (*newNest*).

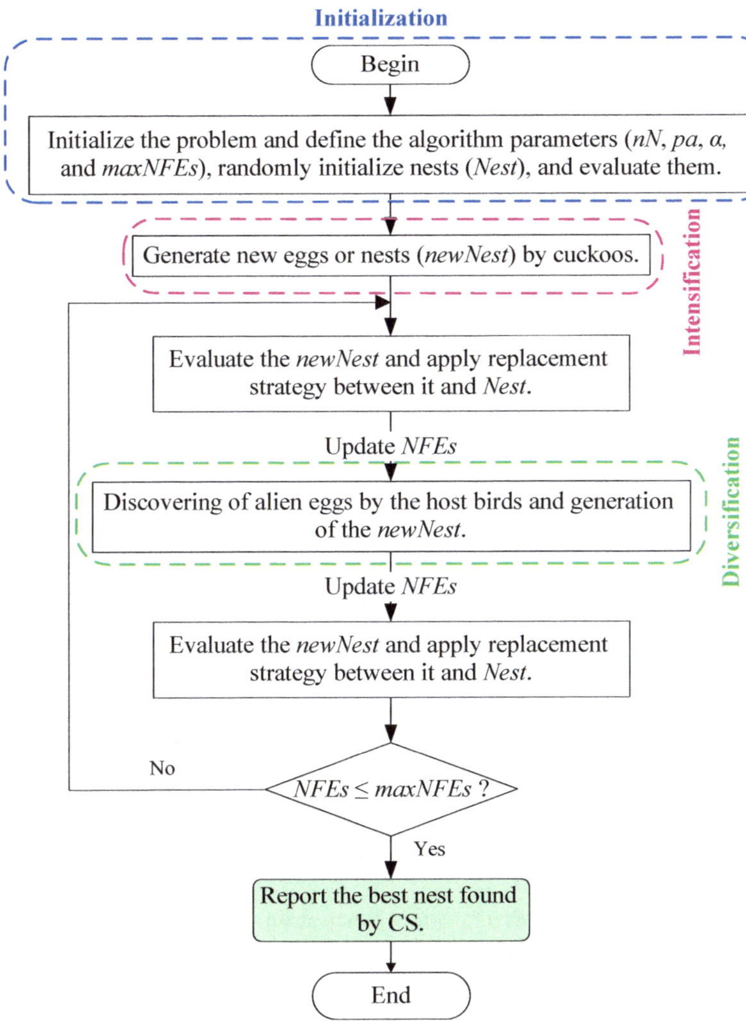

Fig. 7.1 Flowchart of the CS algorithm

Evaluate the *newNest* and apply replacement strategy between old and new eggs.
Update NFEs.
Monitor the best nest (*bestNest*).

end While

7.3 MATLAB Code for Cuckoo Search (CS) Algorithm

According to the previous section, four functions are considered for coding the algorithm. First, these functions are presented with the order of their recall in the algorithm, and at the end, the algorithm is coded. The required comments are presented. Note that the percent sign (%) represents a comment.

The first function is the objective function evaluation (*fobj*) which was presented in Chap. 2.

The second function is named as the *Cuckoo* function to generate new eggs from the current set of solutions toward the known best solution until yet (*bestNest*) in combination with the Lévy flight. The input arguments are the current set of nests matrix (*Nest*), the best nest found by the algorithm until yet (*bestNest*), and the step size parameter (α). The output argument is the new set of produced eggs by the cuckoos (*newNest*). The *Cuckoo* function is coded as follows.

```
% Cuckoo breeding.
function newNest=Cuckoo(Nest,bestNest,alfa)

% Determine random walk based on the Lévy flights (S)
beta=3/2;
sigma=(gamma(1+beta)*sin(pi*beta/2)/(gamma((1+beta)/2)*beta*2^((beta-
1)/2)))^(1/beta);
u=randn(size(Nest)).*sigma;
v=randn(size(Nest));
S=u./abs(v).^(1/beta);

% Determine the step size toward the best solution in combination with the
Lévy flight.
stepsize=randn(size(Nest)).*alfa.*S.*(Nest-
repmat(bestNest,[size(Nest,1),1]));

newNest=Nest+stepsize;
```

The third function is the *Replacement* function in which the old eggs (*Nest*) are evaluated, compared, and replaced with the newly generated ones (*newNest*) in a simple greedy manner so that the egg with smaller penalized objective function (*pfit*) will be preferred. The *fobj* function is called in the nested form within this function. The input arguments are the new set of nests or eggs (*newNest*), old set of eggs (*Nest*) and its corresponding objective (*Fit*) and penalized objective function (*PFit*) vectors, and the lower (*Lb*) and upper bound (*Ub*) of design variables. The output arguments are the updated set of nests and its corresponding objective and penalized objective function vectors. The *Replacement* function is coded as follows.

```
% Evaluating and Updating eggs by comparing the old and new ones.

function [Nest,Fit,PFit]=Replacemnet(Nest,newNest,Fit,PFit,Lb,Ub)

for i=1:size(Nest,1),
    [X,fit,pfit]=fobj(newNest(i,:),Lb,Ub);
    if pfit<=PFit(i)
        Nest(i,:)=X;
        Fit(i)=fit;
        PFit(i)=pfit;
    end
end
```

The fourth function is the alien eggs discovery function by the host birds and named as *Host*. Input arguments are the current set of solutions (*Nest*) and the parameter of the discovery rate (*pa*). The output argument is the newly set of nests (*newNest*) generated by the host birds. The *Host* function is coded as follows.

```
% Alien eggs discovery by the host birds.
function newNest=Host(Nest,pa)

% Form the discovering probability matrix (P).
P=rand(size(Nest))>pa;

% Generate the random permutation based step size.
stepsize=rand(size(Nest)).*(Nest(randperm(size(Nest,1)),:)-
Nest(randperm(size(Nest,1)),:));

newNest=Nest+stepsize.*P;
```

The CS algorithm is coded in the following composed of three parts: initialization, algorithm cyclic body, and monitoring the results. It should be noted that the *Replacement* function will be called literally after both search phases of the algorithm and the number of objective function evaluations of the algorithm (*NFEs*) should be updated. As it was mentioned, based on Eq. (7.5), the best nest will be kept unchanged in the cuckoos breeding search phase, and it does not need to be evaluated. However, for simplicity it will also be evaluated in the *Cuckoos* function. Therefore, in both search phases, *NFEs* will increase by the number of nests (*nN*).

```
clc
clear
%% Initialization

% Define the properties of COP (tension/compression spring design problem).
nV=3; % Number of design variables.
Lb=[0.05 0.25 2]; % Lower bounds of design variables.
Ub=[2 1.3 15]; % Upper bounds of design variables.

% Define parameters of the CS algorithm.
nN=20; % Number of Nests.
pa=.2; % Discovery rate of alien eggs.
alfa=1; % Step size parameter.
maxNFEs=20000; % Maximum number of Objective Function Evaluations.

%Generate random initial solutions or the eggs of host birds.
for i=1:nN
    Nest(i,:)=Lb+(Ub-Lb).*rand(1,nV); % Nests matrix or matrix of the
initial candidate solutions or the initial population.
end

% Evaluate initial population (Nest) calling the fobj function constructed
in the second chapter and form its corresponding vectors of objective
function (Fit) and penalized objective function (PFit). It should be noted
that the design vectors all are inside the search space.
for i=1:nN
    [X,fit,pfit]=fobj(Nest(i,:),Lb,Ub);
    Nest(i,:)=X;
    Fit(i)=fit;
    PFit(i)=pfit;
end

%Monitor the best candidate solution (bestNest) and its corresponding
objective function (minFit) and penalized objective function (minPFit).
[minPFit,m]=min(PFit);
minFit=Fit(m);
bestNest=Nest(m,:);

%% Algorithm Body

NFEs=0; % Current number of Objective Function Evaluations used by the
algorithm until yet.
NITs=0; % Number of algorithm iterations

while NFEs<maxNFEs
    NITs=NITs+1; % Update the number of algorithm iterations.

    % Cuckoo breeding.
    newNest=Cuckoo(Nest,bestNest,alfa);

    % Replace the eggs of host birds with the newly generated ones by
cuckoos if the newest ones are better. Notice that the fobj function is
called in the following replacement function in nested form. Hence the
newly generated eggs will be corrected and evaluated.
    [Nest,Fit,PFit]=Replacemnet(Nest,newNest,Fit,PFit,Lb,Ub);

    % Update the number of objective function evaluations used by the
algorithm until yet.
    NFEs=NFEs+nN;
```

```
    % Alien eggs discovery by the host birds
    newNest=Host(Nest,pa);

    % Replace the eggs of host birds with the newly generated ones by
cuckoos if the newest ones are better. Notice that the fobj function is
called in the following replacement function in nested form. Hence the
newly generated eggs will be corrected and evaluated.
    [Nest,Fit,PFit]=Replacemnet(Nest,newNest,Fit,PFit,Lb,Ub);

    % Monitor the best candidate solution (bestNest) and its corresponding
penalized objective function (minPFit) and objective function (minFit).
    [minPFit,m]=min(PFit);
    minFit=Fit(m);
    bestNest=Nest(m,:);

    % Display desired information of the iteration.
    disp(['NITs= ' num2str(NITs) '; minFit = ' num2str(minFit) '; minPFit
= ' num2str(minPFit)]);

    % Save the required results for post processing and visualization of
algorithm performance.
    output1(NITs,:)=[minFit,minPFit,NFEs];
    output2(NITs,:)=[min(PFit),max(PFit),mean(PFit)];
    output3(NITs,:)=[bestNest,NFEs];
end

%% Monitoring the results
figure;
plot((1:1:NITs),output2(:,1),'g',(1:1:NITs),output2(:,2),'r--
',(1:1:NITs),output2(:,3),'b-.')
legend('min','max','mean');
xlabel('NITs');
ylabel('PFit');
```

7.4 Experimental Evaluation

The CS has four parameters: number of nests (*nN*), step size controlling parameter (α), discovering probability (*pa*), and maximum number of objective function evaluations as the stopping criteria (*maxNFEs*). Considering a large enough value for *maxNFEs* to ensure convergence of the algorithms is essential. This parameter is almost problem dependent and should be considered larger for complicated and larger problems. It is considered here equal to 20,000.

Convergence histories of the algorithm considering different values for the *nN* are monitored in the Fig. 7.2. The values of *pa* and α parameters are considered equal to 0.2 and 1.0, respectively. The *Y* axis is limited for more clarity. CS needs at least 4 number of nest for solving this problem. For values less than 4, the algorithm shows no convergence and is not included in the figure. As it is clear, larger values of the *nN* result in lower convergence speed and accuracy. Simulation results show that values between 7 and 20 lead to the best performance of the algorithm in this COP.

To investigate the effect of the *pa* parameter on the algorithm performance, considering different values of *pa*, CS started from a fixed initial population with

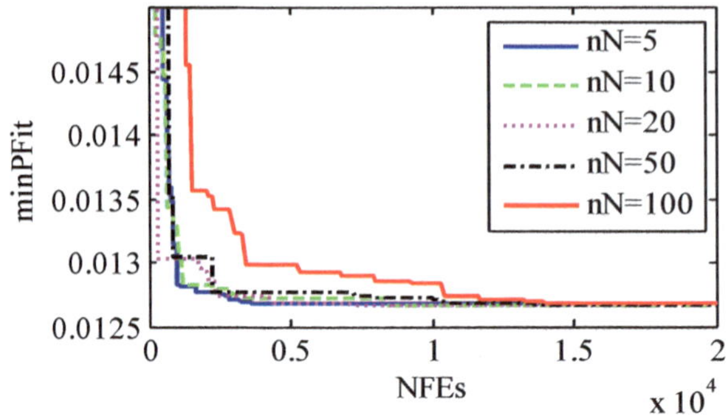

Fig. 7.2 Convergence histories of the algorithm considering different values for *nN*

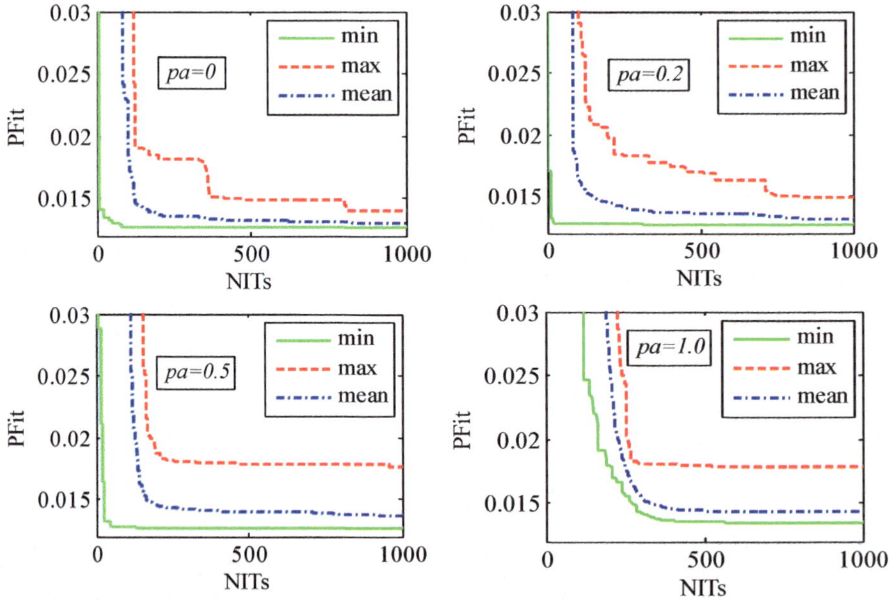

Fig. 7.3 Convergence histories of the minimum, maximum, and mean of the algorithm population considering different values for *pa*

nN and α equal to 20 and 1.0, respectively. Convergence histories for the best, worst, and the average of penalized objective function of the population are depicted in Fig. 7.3. The *Y* axis is limited due to more clarity. As it is clear, considering values between 0.1 and 0.5 results in better performance of the algorithm. It should be noted that 0 value of *pa* stands to discovering of cuckoo eggs and does not work at all and 1 value of *pa* stands that it works completely for all components of all eggs.

Such previous convergence histories are also monitored for CS considering different values for the α parameter in Fig. 7.4. CS applied from a fixed initial

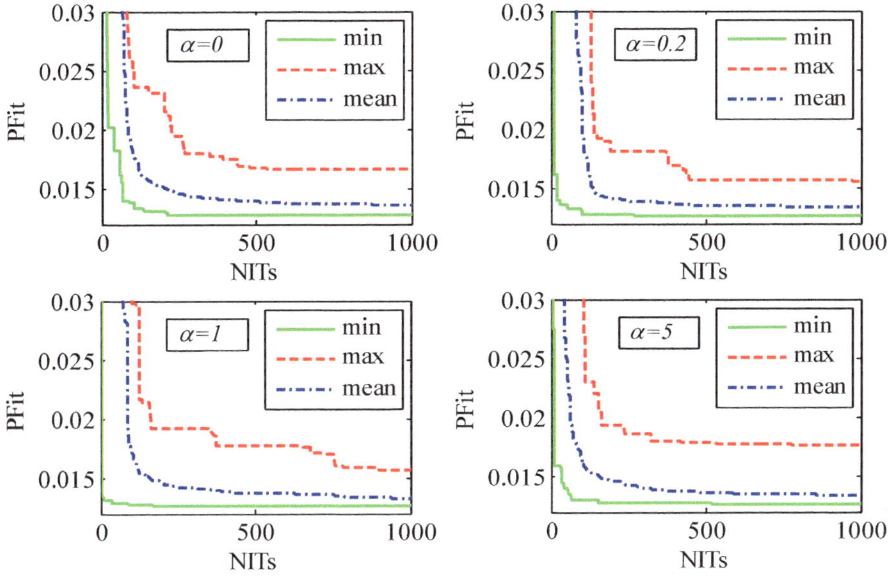

Fig. 7.4 Convergence histories of the minimum, maximum, and mean of the algorithm population considering different values for α

population considering *nN* and *pa* equal to 20 and 0.2, respectively. The Y axis is limited due to more clarity. It should be noted that considering zero value for α means that the cuckoo breeding phase in combination with the Lévy flight does not work and CS searches for the optimum randomly within the second phase. Interesting point is that in this case CS can converge to the near optimum results. Considering the α parameter equal to 1 results in the best performance of the algorithm.

References

1. Yang XS, Deb S (2008) Nature-inspired metaheuristic algorithms. Luniver Press, Bristol
2. Yang XS, Deb S (2010) Engineering optimisation by cuckoo search. Int J Math Model Numer Optim 1:330–343
3. Kaveh A, Bakhshpoori T, Afshari E (2011) An optimization-based comparative study of double layer grids with two different configurations using cuckoo search algorithm. Int J Optim Civ Eng 1:507–520
4. Kaveh A, Bakhshpoori T, Ashoory M (2012) An efficient optimization procedure based on cuckoo search algorithm for practical design of steel structures. Int J Optim Civ Eng 2:1–14

Chapter 8
Charged System Search Algorithm

8.1 Introduction

Charged System Search (CSS) algorithm was developed by Kaveh and Talatahari [1, 2] as an efficient population-based metaheuristic using some principles from physics and mechanics and was applied successfully to various types of structural optimization problems [3–7]. CSS utilizes the governing Coulomb laws from electrostatics and the Newtonian laws of mechanics. In this algorithm each agent is a charged particle with a predetermined radius. The charge of magnitude of particles is considered based on their quality. Each particle creates an electric field, which exerts a force on other electrically charged objects. Therefore, charged particles can affect each other based on their fitness values and their separation distance. The quantity of the resultant force is determined by using the electrostatics laws, and the quality of the movement is determined using Newtonian mechanics laws.

8.2 Formulation and Framework of the Charged System Search Algorithm

In the following, the first subsection gives a brief overview of the electrical and mechanics laws which constitute the essence of the CSS [1]. Then the formulation, pseudo code, and flowchart of the algorithm are presented in the subsequent subsection [2].

A. Kaveh, T. Bakhshpoori, *Metaheuristics: Outlines, MATLAB Codes and Examples*, https://doi.org/10.1007/978-3-030-04067-3_8

8.2.1 Electrical and Mechanics Laws

Electrical Laws
Consider two charged particles (i and j) with a total positive charge q_i and q_j, and positions of r_i and r_j, respectively. Based on the Coulomb law, the electric force between them is defined as:

$$F_{ij} = k_e \frac{q_i q_j}{r_{ij}^2} \frac{r_i - r_j}{\|r_i - r_j\|} \tag{8.1}$$

in which k_e is the Coulomb constant and r_{ij} is the distance between two charges.

Consider an insulating solid sphere of radius a which has a uniform volume charge density and carries a total charge of magnitude q_i. The magnitude of the electric force at a point outside the sphere is defined as Eq. (8.1), while this force can be obtained using Gauss's law at a point inside the sphere as:

$$F_{ij} = k_e \frac{q_i q_j}{a^3} r_{ij} \frac{r_i - r_j}{\|r_i - r_j\|} \tag{8.2}$$

It should be noted that the electric field inside the sphere varies linearly with the distance and is inversely proportional to the square of the distance. We will see how CSS uses these two points for balancing between exploration and exploitation.

In order to calculate the electric force on a charge (q_j) at a point (r_j) due to a set of point charges, the principle of superposition is applied to electric forces as:

$$F_j = \sum_{i=1,\quad i \neq j}^{nCP} F_{ij} \tag{8.3}$$

where nCP is the total number of charged particles and F_{ij} is equal to:

$$F_{ij} = \begin{cases} \dfrac{k_e q_i}{a^3} r_{ij} \dfrac{r_i - r_j}{\|r_i - r_j\|} & \text{if } r_{ij} < a \\ \dfrac{k_e q_i}{r_{ij}^2} \dfrac{r_i - r_j}{\|r_i - r_j\|} & \text{if } r_{ij} \geq a \end{cases} \tag{8.4}$$

Therefore, the resulted electric force can be obtained as:

$$F_j = k_e q_j \sum_{i=1,\quad i\neq j}^{nCP} F_{ij} \left(\frac{q_i}{a^3} r_{ij} \cdot i_1 + \frac{q_i}{r_{ij}^2} \cdot i_2 \right) \frac{r_i - r_j}{r_i - r_j}$$
$$i_1 = 1, i_2 = 0 \quad \text{if } r_{ij} < a$$
$$i_1 = 0, i_2 = 1 \quad \text{if } r_{ij} \geq a \tag{8.5}$$

Newtonian Mechanics Laws

In Newtonian mechanics motion studies, the displacement of a particle that moves from an initial position r_{old} to a new position r_{new} is given by:

$$\Delta r = r_{new} - r_{old} \tag{8.6}$$

and the slope of tangent line of the particle position represents the velocity of this particle as:

$$V = \frac{r_{new} - r_{old}}{t_{new} - t_{old}} = \frac{r_{new} - r_{old}}{\Delta t} \tag{8.7}$$

and the particle is said to be accelerated when the velocity changes with time, and the acceleration of the particle is defined as:

$$a = \frac{V_{new} - V_{old}}{\Delta t} \tag{8.8}$$

Using the above equations, the displacement of any object as a function of time is obtained as:

$$r_{new} = \frac{1}{2} a \Delta t^2 + V_{old} \cdot \Delta t + r_{old} \tag{8.9}$$

Also according to Newton's second law, we have

$$F = m \cdot a \tag{8.10}$$

where m is the mass of the object. Substituting Eq. (8.10) in Eq. (8.9), we have

$$r_{new} = \frac{1}{2} \frac{F}{m} \Delta t^2 + V_{old} \cdot \Delta t + r_{old} \tag{8.11}$$

8.2.2 Charged System Search Algorithm

Looking at the electrical and Newtonian laws overviewed at the previous subsection, there is an analogy between these laws and a population-based metaheuristic. Each candidate solution is a charged particle in the search space. The value of charge and mass of particles is proportional to their objective function, so that a less objective function value leads to the greater charge and mass. Analogously, in each iteration, transitions of particles can be induced by electric fields leading to particle-particle electrostatic interactions with the aim of attracting or repelling the particles toward the optimum position. The quantity of the resultant electric force and the quality of the movement will be determined by using the electrostatics laws and Newtonian mechanics laws, respectively. At the following, the CSS algorithm is presented in the form of nine rules:

Rule 1

CSS considers a set of charged particles (*CP*) with the total number of *nCP*, where each charged particle has a penalized objective function of *PFit(i)* and charge of magnitude of *q(i)* defined considering the quality of its solution as:

$$q(i) = \frac{PFit(i) - \max(PFit)}{\min(PFit) - \max(PFit)}; \; i = 1, 2, 3, \ldots, nCP \tag{8.12}$$

Here min and max are the minimum and maximum functions that return the penalized objective function of the best and the worst of the current particles, respectively. It should be noted that the current best and worst particles can be considered as the so far best and worst particles, the CSS created from its first iteration. However, the current form is defined here for simplicity. In this way the best and worst particles will possess a magnitude of charge equal to 1 and 0, respectively. The separation distance *r(ij)* between two charged particles is defined as follows:

$$r(ij) = \frac{\|CP(i) - CP(j)\|}{\left\| \frac{CP(i)+CP(j)}{2} - bestCP \right\| + \varepsilon} \tag{8.13}$$

where *CP(i)* and *CP(j)* are the positions of the *i*th and *j*th charged particles, respectively, *bestCP* is the position of the best current charged particle with the minimum value of the penalized objective function, and ε is a small positive number to avoid singularities.

Rule 2

The initial positions of charged particles are determined randomly in the search space, and the initial velocities (*V*) of charged particles are assumed to be 0.

Rule 3
Electric forces between any two charged particles are attractive. This rule increases the exploitation ability of the algorithm. Though it is possible to define repelling force between charged particles as well, which seems to be unnecessary for many problems. When a search space is a noisy domain, having a complete search before converging to a result is necessary; in such conditions the addition of the ability of repelling forces to the algorithm may improve its performance.

Rule 4
Various conditions can be considered related to the kind of the attractive forces. However, considering conditions such that all good charged particles can attract bad ones and only some of the bad agents attract good agents can be efficient. This can be considered using the following probability function:

$$p(ij) = \begin{cases} 1 & \dfrac{PFit(i) - \min(PFit)}{PFit(j) - PFit(i)} > rand \text{ or } PFit(j) > PFit(i) \\ 0 & \text{else} \end{cases} \tag{8.14}$$

According to the above equation, when a good agent attracts a bad one, the exploitation ability for the algorithm is provided, and vice versa if a bad particle attracts a good one, the exploration is provided. When a charged particle moves toward a good agent, it improves its performance, and so the self-adaptation principle is guaranteed. Moving a good charged particle toward a bad one may cause losing the previous good solution or at least increasing the computational cost to find a good solution. To resolve this problem, a memory which saves the best-so-far solutions can be considered which will be presented as the seventh rule.

Rule 5
The value of the resultant electrical force affecting a charged particle is determined using Eq. (8.15):

$$\begin{gathered} f(j) = q(j) \sum_{i,\ i \neq j}^{nCP} \left(\frac{q(i)}{a^3} r(ij) \cdot i_1 + \frac{q(i)}{r(ij)^2} \cdot i_2 \right) p(ij)(CP(i) - CP(j)) \\ \left\langle \begin{array}{ll} i_1 = 1, i_2 = 0 & \text{if } r_{ij} < a \\ i_1 = 0, i_2 = 1 & \text{if } r_{ij} \geq a \end{array} \right. \end{gathered} \tag{8.15}$$

where f_j is the resultant force acting on the jth charged particle. CSS considers each agent as a charged sphere with radius a having a uniform volume charge density. a can be set to unity; however, for more complex examples, the appropriate value for a must be defined considering the size of the search space. The following equation can be utilized as a general formula:

$$a = 0.1 \times \max(Lb - Ub) \tag{8.16}$$

in which *Lb* and *Ub* are the practical range of design variables of the optimization problem at hand.

According to this rule, in the early iterations where the agents are far from each other, the magnitude of the resultant force acting on a particle is inversely proportional to the square of the separation between the particles. Thus the exploration power in this condition is high due to performing more searches in the early iterations. It is necessary to increase the exploitation of the algorithm and to decrease the exploration gradually. After a number of searches where charged particles are collected in a small space and the distance between them becomes small, say 0.1, then the resultant force becomes proportional to the separation distance of the particles instead of being inversely proportional to the square of the separation distance. If the first case ($f(ij) \propto 1/r(ij)^2$) is used for $r(ij) = 0.1$, we have $f(ij) = 100 \times k_e q(i)q(j)$ that is a large value, compared to a force acting on a particle at $r(ij) = 2$ ($f(ij) = 0.25 \times k_e q(i)q(j)$), and this great force causes particles to get farther from each other instead of getting nearer, while the second one ($f(ij) \propto r(ij)$) guarantees that a convergence will happen. Therefore, the parameter *a* separates the global search phase and the local search phase, i.e., when majority of the agents are collected in a space with radius *a*, the global search is finished, and the optimizing process is continued by improving the previous results, and thus the local search starts.

Besides controlling the balance between the exploration and the exploitation, this rule guarantees the competition step of the algorithm. Since the resultant force is proportional to the magnitude of the charge, a better fitness (great $q(i)$) can create a stronger attracting force, so the tendency to move toward a good particle becomes more than toward a bad particle.

Rule 6

The new position (*newCP*) and velocity (*newV*) of each charged particle is determined considering Eqs. (8.17) and (8.18), as follows:

$$newCP(j) = rand \cdot k_a \cdot \frac{f(j)}{m(j)} \cdot \Delta t^2 + rand \cdot k_v \cdot V(j) \cdot \Delta t + CP(j) \tag{8.17}$$

$$newV(j) = \frac{newCP(j) - CP(j)}{\Delta t} \tag{8.18}$$

where k_a and k_v are the acceleration and velocity control coefficients, respectively, and rand is a random number uniformly distributed in the range of (0, 1), $m(j)$ is the mass of the charged particle which is equal to $q(j)$, and Δt is the time step and is set to unit. Notice the first component of Eq. (8.17) ($rand \cdot k_a \cdot \frac{f(j)}{m(j)} \cdot \Delta t^2$). It should be noted that the resultant electrical force ($f(j)$) affecting *j*th charged particle Eq. (8.5) has the value $q(j)$ in the numerator which is equal to $m(j)$ in the denominator. Then the $q(j)$ and $m(j)$ will be eliminated. In this way the singularity problem for the worst particle with the charge and mass equal to 0 will be avoided.

The effect of the pervious velocity and the resultant force acting on a charged particle can be decreased or increased based on the values of the k_v and k_a, respectively. Excessive search in the early iterations may improve the exploration ability; however, it must decrease gradually, as described before. Since k_a is the parameter related to the attracting forces, selecting a large value for this parameter may cause a fast convergence, and vice versa a small value can increase the computational time. In fact, k_a is a control parameter of the exploitation. Therefore, choosing an incremental function can improve the performance of the algorithm. Also, the direction of the pervious velocity of a charged particle is not necessarily the same as the resultant force. Thus, it can be concluded that the velocity coefficient k_v controls the exploration process, and therefore a decreasing function can be considered. Thus, k_v and k_a are defined as:

$$\begin{aligned} k_v &= 0.5(1 - NFEs/maxNFEs) \\ k_a &= 0.5(1 + NFEs/maxNFEs) \end{aligned} \tag{8.19}$$

where *NFEs* is the number of objective function evaluations CSS used and *maxNFEs* is the maximum number of objective function evaluations defined as the stopping criteria of the algorithm. In this way, the balance between the exploration and fast rate of convergence is saved.

Rule 7

Considering a memory (*CP-M*) which saves the best charged particles and their related objective function (*Fit-M*) and penalized objective function (*PFit-M*) values can improve the algorithm performance without increasing the computational cost. To fulfill this aim, charged memory is utilized to save a number of the best-so-far solutions. The size of the charged memory is taken as *nCP*/4. Charged memory can be used for position correction of the agent's exit from the search space. Another benefit of the memory can be a guidance of the current charged particles. In other words, the vectors stored in the *CP-M* can attract current charged particles according to Eq. (8.15). Instead, it is assumed that the same number of the current worst particles cannot attract the others. Here only the first usage of charged memory is benefited.

Rule 8

There are two major problems related to many metaheuristic algorithms; the first problem is the balance between exploration and exploitation in the beginning, during, and at the end of the search process, and the second is how to deal with an agent violating the limits of the variables.

The first problem is solved naturally through the application of the above-stated rules; however, in order to solve the second problem, one of the simplest approaches is utilizing the nearest limit values for the violated variable. Alternatively, one can force the violating particle to return to its previous position, or one can reduce the maximum value of the velocity to allow fewer particles to violate the variable boundaries. Although these approaches are simple, they are not sufficiently efficient and may lead to reduce the exploration of the search space. This problem has previously been addressed and solved using the harmony search-based handling approach [8]. According to this mechanism, any component of the solution vector violating the variable boundaries can be regenerated from the *CP-M* as:

$$CP(i,j) = \begin{cases} \text{w.p. } cmcr & \Rightarrow \text{ select a new value for a variable from } CP - M, \\ & \Rightarrow \text{ w.p. } (1 - par) \text{ do nothing,} \\ & \Rightarrow \text{ w.p. } par \text{ choose a neighboring value,} \\ \text{w.p. } (1 - cmcr) & \Rightarrow \text{ select a new value randomly,} \end{cases} \tag{8.20}$$

where "w.p." is the abbreviation for "with the probability," *CP(i, j)* is the *j*th component of the *i*th charged particle, *cmcr* is the charged memory considering rate varying between 0 and 1 and sets the rate of choosing a value in the new vector from the historic values stored in the *CP-M*, and (1 − *cmcr*) sets the rate of randomly choosing one value from the possible range of values. The pitch-adjusting process is performed only after a value which is chosen from *CP-M*. The value (1 − *par*) sets the rate of doing nothing, and *par* sets the rate of choosing a value from neighboring the best charged particle or the particles saved in memory. For choosing a value from neighboring the best charged particle or the particles saved in memory, for continuous search space simply a randomly generated step size can be used ($\pm bw \times$ rand). For discrete search space, the reader can refer to the [8].

Rule 9

Maximum number of objective function evaluations (*maxNFEs*) is considered as the terminating criterion.

The pseudo code of CSS is provided as follows, and the flowchart is illustrated in Fig. 8.1.

The pseudo code of the CSS algorithm for solving COPs:

Define the algorithm parameters: *nCP*, *cmcr*, *par*, *bw*, and *maxNFEs*.

Generate random initial solutions or charged particles (*CP*) with random positions and their associated velocities which are 0.

Evaluate initial charged particle matrix (*CP*); form its corresponding vectors of objective function (*Fit*) and penalized objective function (*PFit*).

Form the charged particles memory matrix (*CP_M*) and its corresponding vectors of objective function memory (*Fit_M*) and penalized objective function memory (*PFit_M*).

While *NFEs*<*maxNFEs*

Update the number of algorithm iterations (*NITs*).

Determine the charge of magnitude vector (*q*), the separation distance (*r*), and the resultant electrical force (*f*) matrices of charged particles based on the Coulomb and Gaussian laws.

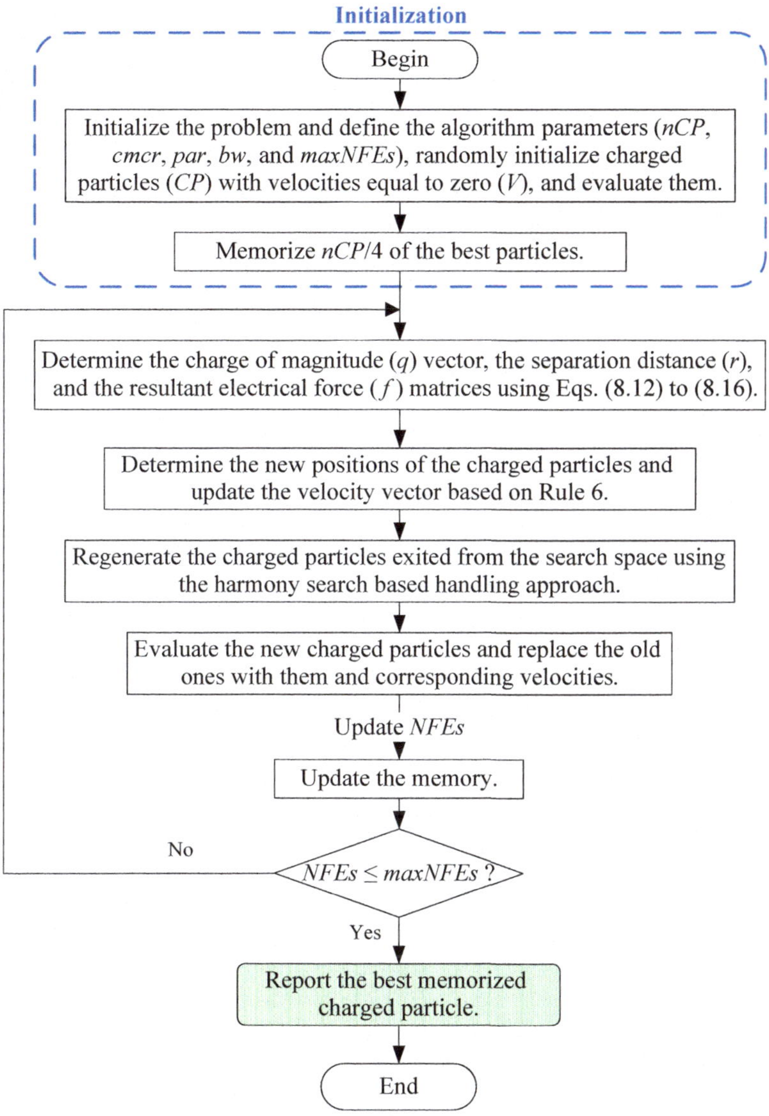

Fig. 8.1 Flowchart of the CSS algorithm

Determine the new positions of the charged particles and update the velocity vector based on the Newtonian laws.
Regenerate the charged particles exited from the search space using the harmony search based handling approach.
Evaluate the new charged particles.
Update *NFEs*.

Replace the old charged particles and their corresponding velocities with the newly generated ones.

Update the charged particles memory matrix (*CP_M*) and its corresponding vectors of objective function memory (*Fit_M*) and penalized objective function memory (*PFit_M*).

Monitor the best memorized candidate solution (*bestCP*).

end While

8.3 MATLAB Code for Charged System Search (CSS) Algorithm

According to the previous section, five functions are considered for coding the algorithm. First, these functions are presented with the order of their recall in the algorithm, and at the end, the algorithm is coded. The required comments are presented. Note that the percent sign (%) represents a comment.

The first function is the objective function evaluation (*fobj*) which was presented in the Chap. 2.

The second function named as the *Coulomb_Laws* determines the charge of magnitude vector (q), the separation distance (r), and the resultant electrical force (f) matrices of charged particles based on the Coulomb and Gaussian laws. The input arguments are the current set of charged particle matrix (*CP*), penalized objective function vector (*PFit*), and the radius of charged particles (a). The output arguments are the q vector and r and f matrices. The *Coulomb_Laws* function is coded as it follows. Notice that the attraction probability function (p) should be determined to use in the calculation of the resultant electrical force.

```
% Determine the charge of magnitude (q) vector, the separation distance (r)
matrix, and the resultant electrical force (f) matrix of charged particles
based on the Coulomb and the Gaussian laws.
function [q,r,f]=Coulomb_Laws(CP,PFit,a)

nCP=size(CP,1);

for i=1:nCP
    q(i)=(PFit(i)-max(PFit))/(min(PFit)-max(PFit));
end

[value,index]=min(PFit);
for i=1:nCP
    for j=1:nCP
        r(i,j)=(norm(CP(i,:)-CP(j,:)))/(norm(((CP(i,:)+CP(j,:))/2)-
CP(index,:))+0.000001);
    end
end

% Determine the attraction probability function (p).
for i=1:nCP
    for j=1:nCP
        if          (((PFit(i)-min(PFit))/(PFit(j)-PFit(i)))>rand)          ||
(PFit(j)>PFit(i))
            p(i,j)=1;
        else
            p(i,j)=0;
        end
    end
end

for j=1:nCP
    fhelp=zeros(1,size(CP,2));
    for i=1:nCP
        if i~=j
            if r(i,j)<a
                fhelp=fhelp+((q(i)/(a^3))*r(i,j))*p(i,j)*(CP(i,:)-CP(j,:));
            else
                fhelp=fhelp+(q(i)/((r(i,j))^2))*p(i,j)*(CP(i,:)-CP(j,:));
            end
        end
    end
    f(j,:)=fhelp';
end
```

The third function is the *Newtonian_Laws* in which the new positions of the charged particles are calculated and the corresponding velocity vectors updated based on the Newtonian laws. The input arguments are the *CP*, *V*, and *f* matrices, *deltaT* as the time step which is set to unit, *NFEs*, and *maxNFEs*. The output arguments are the updated set of charged particles (*newCP*) and their corresponding velocities (*newV*). The *Newtonian_Laws* function is coded as it follows:

```
% Determine the new positions of the charged particles and update the
velocity vector based on the Newtonian laws.

function [newCP,newV]=Newtonian_Laws(CP,V,f,deltaT,NFEs,maxNFEs)

kv=2*(1-NFEs/maxNFEs);ka=0.8*(1+NFEs/maxNFEs);
for i=1:size(CP,1)
    newCP(i,:)=ka*rand*f(i,:)*deltaT^2+kv*rand*V(i,:)*deltaT+CP(i,:);
    newV(i,:)=(newCP(i,:)-CP(i,:))/deltaT;
end
```

The fourth function is regenerating function of the charged particles which exited from the search space using the harmony search-based handling approach named as *Harmony*. Input arguments are the newly generated (*newCP*) set of charged particles, the charged particles memory matrix (*CP_M*), the parameters of harmony search-based handling approach (*cmcr*, *par*, *bw*), and lower (*Lb*) and upper (*Ub*) bounds of design variables. The output argument is the corrected set of particles that swerves the side limits. The *Harmony* function is coded as it follows:

```
% If a charged particle swerves the side limits correct its position using
the Harmony search based handling approach.

function [newCP]=Harmony(newCP,CP_M,cmcr,par,bw,Lb,Ub)

for i=1:size(newCP,1)
    for j=1:size(newCP,2)
        if (newCP(i,j)<Lb(j)) || (newCP(i,j)>Ub(j))
            if rand<cmcr
                newCP(i,j)=CP_M(ceil(rand*(size(newCP,1)/4)),j);
                if rand<par
                    if rand<.5
                        newCP(i,j)=newCP(i,j)+bw*rand;
                    else
                        newCP(i,j)=newCP(i,j)-bw*rand;
                    end
                end
            else
                newCP(i,j)=Lb(j)+(Ub(j)-Lb(j))*rand;
            end
        end
    end
end
```

The fifth function is the *Memory* function by which the charged particles memory matrix (*CP_M*) and its corresponding vectors of objective function memory (*Fit_M*) and penalized objective function memory (*PFit_M*) are updated. The input arguments are *CP*, *Fit*, *PFit*, *CP_M*, *Fit_M*, *PFit_M* and the output arguments are *CP_M*, *Fit_M*, *PFit_M*. The *Memory* function is coded in the following. It should be noted that in the initialization phase, the memorizing should take place firstly. However, the *Memory* function will not be called for memorizing in the initialization phase because of its simplicity in this phase.

```
% Update the charged particles memory matrix(CP_M) and its corresponding
vectors of objective function memory (Fit_M) and penalized objective
function memory (PFit_M).

function [CP_M,Fit_M,PFit_M]=Memory(CP,Fit,PFit,CP_M,Fit_M,PFit_M)
nCP=size(CP,1);

for i=1:nCP
    for j=1:nCP/4
        if PFit(i)<PFit_M(j)
            CP_M(j,:)=CP(i,:);
            Fit_M(j)=Fit(i);
            PFit_M(j)=PFit(i);
            break
        end
    end
end
```

The CSS algorithm is coded in the following and composed of three parts: initialization, algorithm cyclic body, and monitoring the results.

```
clc
clear
%% Initialization

% Define the properties of COP (tension/compression spring design problem).
nV=3; % Number of design variables.
Lb=[0.05 0.25 2]; % Lower bounds of design variables.
Ub=[2 1.3 15]; % Upper bounds of design variables.

% Define the parameters of CSS algorithm.
nCP=20; % Number of Charged Particles.
a=0.1*max(Ub-Lb); % Radios of charged particles.
deltaT=1; % Time step used in the Newtonian laws.
cmcr=0.95;par=0.1;bw=.1; % Parameters for regenerating the CPs exited from
the search space using the harmony search based handling approach
maxNFEs=20000; % Maximum Number of Objective Function Evaluations.

% Generate random initial solutions.
for i=1:nCP
    CP(i,:)=Lb+(Ub-Lb).*rand(1,nV); % Charged particles matrix or matrix of
the initial candidate solutions or the initial population.
end

% Form the initial velocity vector of charged particles (V) which are
assumed to be zero.
V=zeros(nCP,nV);

% Evaluate initial population (CP) calling the fobj function constructed in
the second chapter and form its corresponding vectors of objective function
(Fit) and penalized objective function (PFit). It should be noted that the
design vectors all are inside the search space.
for i=1:nCP
    [X,fit,pfit]=fobj(CP(i,:),Lb,Ub);
    CP(i,:)=X;
    Fit(i)=fit;
    PFit(i)=pfit;
end

% Form the charged particles memory matrix(CP_M) and its corresponding
vectors of objective function memory (Fit_M) and penalized objective
function memory (PFit_M). should be noted that in the initialization phase
it is not needed to utilize the Memory function and the memories simply can
be produced.
[value,index]=sort(PFit);
CP_M=CP(index(1:nCP/4),:);Fit_M=Fit(index(1:nCP/4));PFit_M=PFit(index(1:nCP
/4));

%% Algorithm Body

NFEs=0; % Current Number of Objective Function Evaluations used by the
algorithm until yet.
NITs=0; % Number of algorithm iterations.

while NFEs<maxNFEs
    NITs=NITs +1; % Update the number of algorithm iterations.

    % Determine the charge of magnitude (q), the separation distance (r),
and the resultant electrical force (f) vectors of CPs based on the Coulomb
laws.
```

```
    [q,r,f]=Coulomb_Laws(CP,PFit,a);

    % Determine the new positions of the charged particles and update the
velocity vector based on the Newtonian laws.
    [newCP,Vnew]=Newtonian_Laws(CP,V,f,deltaT,NFEs,maxNFEs);

    % Regenerate the charged particles exited from the search space using
the harmony search based handling approach.
    [newCP]=Harmony(newCP,CP_M,cmcr,par,bw,Lb,Ub);

    % Evaluate the new CPs calling the fobj function and replace the old
charged particles and their corresponding velocities with the newly
generated ones. It should be noted that the existed dimensions are
corrected before within the Harmony function. However, to do not change the
form of fobj function it is checked again here which is not needed.
    for i=1:nCP
        [X,fit,pfit]=fobj(newCP(i,:),Lb,Ub);
        CP(i,:)=X;
        Fit(i)=fit;
        PFit(i)=pfit;
    end
    NFEs=NFEs+nCP;
    V=Vnew;

    % Update the charged particles memory matrix(CP_M) and its
corresponding vectors of objective function memory (Fit_M) and penalized
objective function memory (PFit_M) utilizing the Memory function.
    [CP_M,Fit_M,PFit_M]=Memory(CP,Fit,PFit,CP_M,Fit_M,PFit_M);

    % Monitor the best memorized candidate solution (bestCP) and its
corresponding objective function (minFit) and penalized objective function
(minPFit).
    minPFit=PFit_M(1);minFit=Fit_M(1);bestCP=CP_M(1,:);


    % Display desired information of the iteration.
    disp(['NITs= ' num2str(NITs) '; minFit = ' num2str(minFit) '; minPFit
= ' num2str(minPFit)]);

    % Save the required results for post processing and visualization of
the algorithm performance.
    output1(NITs,:)=[minFit,minPFit,NFEs];
    output2(NITs,:)=[min(PFit),max(PFit),mean(PFit)];
    output3(NITs,:)=[bestCP,NFEs];
end

%% Monitoring the results.
figure;
plot((1:1:NITs),output2(:,1),'g',(1:1:NITs),output2(:,2),'r--
',(1:1:NITs),output2(:,3),'b-.')
legend('min','max','mean');
xlabel('NITs');
ylabel('PFit');
```

8.4 Experimental Evaluation

The CSS uses the following parameters in its formulation: number of charged particles (*nCP*), radios of charged particles (a), size of memory, time step used in the Newtonian laws (Δt), parameters for regenerating the charged particles which exited from the search space using the harmony search-based handling approach (charged memory considering rate parameter, *cmcr*; pitch-adjusting rate parameter, *par*; and the step size parameter for choosing a value from neighboring of the particles saved in memory, *bw*), the velocity (k_v) and acceleration (k_a) control coefficients used in the Newtonian laws, and maximum number of objective function evaluations as the stopping criteria (*maxNFEs*). It should be noted that all the parameters except the *nCP* and *maxNFEs* and the parameters of the harmony search-based handling approach are formulated based on the properties of problem or other parameters of the algorithm, are defined in Sect. 8.2.2. For example, size of memory is defined as a quarter of *nCP*. However, the formulation for k_v and k_a is needed to be tuned. Based on Eq. (8.19), the velocity (k_v) and acceleration (k_a) control coefficients should be considered so that they incrementally decreased from 0.5 to 0 and increased from 0.5 to 1, respectively. This formulation results in the bad performance of the algorithm. It is found that in this problem considering 2 and 0.8 instead of 0.5 for k_v and k_a, respectively, results in the better performance of the algorithm.

Considering a large enough value for *maxNFEs* to ensure convergence of the algorithms is essential. This parameter is almost problem dependent and should be considered larger for complicated and larger problems. It is considered here equal to 20,000. Convergence histories of the algorithm for a single run considering different values for the *nCP* are monitored in Fig. 8.2. The *Y* axis is limited for more clarity. In this problem the smaller values for *nCP* further endorse the higher accuracy and convergence speed of the algorithm.

According to Rule 5, the parameter a separates the global search and the local search phase. The trend of convergence histories is obtained for different values of a captures this fact as evidenced from Fig. 8.3. It should be noted that in all cases the

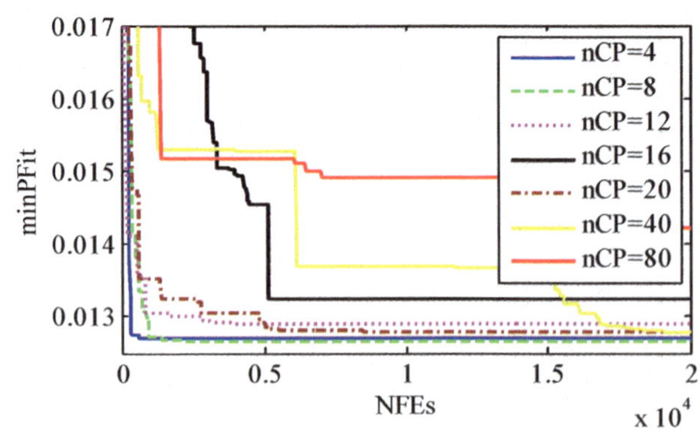

Fig. 8.2 Convergence histories of the algorithm considering different values for *nCP*

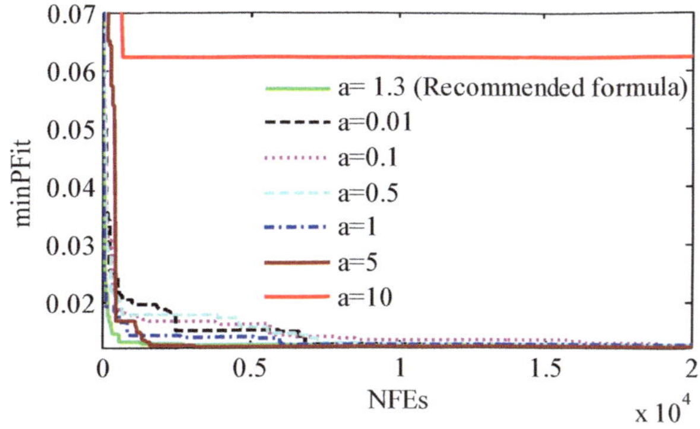

Fig. 8.3 Convergence histories of the algorithm considering different values for *a*

nCP is considered equal to 20 and the algorithm is run starting from a fixed initial population. For a large value of *a*, algorithm performs merely local search, and all particles are converged to and trapped in the local optima. For small values of *a*, CSS performs the search more globally, and the convergence speed is low. As it is clear, the recommended formula results in the best result in both aspects of accuracy and convergence speed and can make a good balance between diversification and intensification.

To investigate the effect of memory size parameter (*nCP*/4) and also the parameters of the harmony search-based handling approach ($cmcr = 0.95$, $par = 0.1$, and $bw = 0.1$) to correct the position of the swerved particles from the search space, experimental studies are made, and convergence histories are monitored. The simulation results show the efficiency of the recommended values.

8.5 Extension to CSS

The Magnetic Charged System Search (MCSS) algorithm is an extended version of the CSS developed by Kaveh et al. [9]. The difference between these two algorithms is that CSS only considers the electric force but MCSS includes magnetic forces besides electric forces. The main structure of the algorithm is the same as the CSS, but a change is made in part of the algorithm where the forces are computed. By using the physical laws on magnetic fields and forces, magnetic forces can be determined. Each solution candidate $\mathbf{X}_i$ called CP (charged particle) contains electrical charge. These CPs produce electric fields and exert electric forces on each other. When a CP moves, it creates a magnetic field in the space, and this magnetic field imposes magnetic forces on other CPs. For further explanation of the MCSS, the reader can refer to [9].

References

1. Kaveh A, Talatahari S (2010) Optimal design of skeletal structures via the charged system search algorithm. Struct Multidiscip Optim 41:893–911
2. Kaveh A, Talatahari S (2010) A novel heuristic optimization method: charged system search. Acta Mech 213(3–4):267–289
3. Kaveh A, Talatahari S (2010) Charged system search for optimum grillage system design using the LRFD-AISC code. J Construct Steel Res 66:767–771
4. Kaveh A, Talatahari S (2011) Geometry and topology optimization of geodesic domes using charged system search. Struct Multidiscip Optim 43:215–229
5. Talatahari S, Kaveh A, Mohajer Rahbari N (2012) Parameter identification of Bouc-Wen model for MR fluid dampers using adaptive charged system search optimization. J Mech Sci Tecnol 26:2523–2534
6. Kaveh A, Talatahari S, Farahmand Azar B (2012) Optimum design of composite open channels using charged system search algorithm. Iran J Sci Technol Trans Civ Eng 36(C1):67
7. Kaveh A, Talatahari S (2012) Charged system search for optimal design of frame structures. Appl Soft Comput 12:382–393
8. Kaveh A, Talatahari S (2009) Particle swarm optimizer, ant colony strategy and harmony search scheme hybridized for optimization of truss structures. Comput Struct 87(5–6):267–283
9. Kaveh A, Motie Share MA, Moslehi M (2013) A new meta-heuristic algorithm for optimization: magnetic charged system search. Acta Mech 224(1):85–107

Chapter 9
Ray Optimization Algorithm

9.1 Introduction

Kaveh and Khayatazad [1] developed the Ray Optimization (RO) algorithm as a novel population-based metaheuristic conceptualized based on Snell's light refraction law when light travels from a lighter medium to a darker medium. RO applied successfully to some of the structural optimization problems [2–5].

Based on Snell's light refraction law, when light travels from a medium to another, it refracts. The refraction depends on (1) the angle between the incident ray and the normal vector of the interface surface of two mediums and (2) the refraction index ratio of two mediums. Its direction changes in a way that gets closer to the normal vector when it passes from a lighter medium to a darker one. This physical behavior is the essence of the RO. The agents of RO are considered as beginning points of rays of light updated in the search space or traveled from a medium to another one based on Snell's light refraction law. Each ray of light is a vector so that its beginning point is the previous position of the agent in the search space, its direction and length is the searching step size in the current iteration, and its end point is the current position of the agent achieved by adding the step size to the beginning point. Considering an effective vector as the normal vector of the interface surface between two mediums and an effective value for the refraction index ratio of two mediums, the refraction vector can be achieved based on Snell's law as the new searching step size. Consequently, the new position of agents are updated to explore the search space and converge to the global or near-global optimum. The current position of the agents as one of the starting or ending points of this vector is inevitable to use the Snell's law. However, the other one should be selected in a way that creates a good balance between exploration and exploitation. RO considered effectively the normal vector so that it starts from a point determined based on the individual and collective knowledge of agents and ends in the current position of agents. RO starts from a randomly generated initial candidate solutions and random initial search step sizes. These are the rays of light that travel from a

A. Kaveh, T. Bakhshpoori, *Metaheuristics: Outlines, MATLAB Codes and Examples*, https://doi.org/10.1007/978-3-030-04067-3_9

medium to another in the cyclic body of the algorithm. In fact, RO aims to improve the quality of the solutions by refracting the rays toward the promising points obtained based on the best-known solution by each agent and all of them.

Kaveh et al. [6] developed an Improved Ray Optimization (IRO) algorithm to enhance the accuracy and convergence rate of the RO. The framework of IRO is based on the original RO. However, here the IRO is coded because of its efficiency and simplicity over RO algorithm.

9.2 Formulation and Framework of the RO and IRO

In the following, firstly, Snell's light refraction law as the essence of RO is presented. The second subsection captures the analogy between this law and a metaheuristic. Then the concepts of the RO algorithm are presented. In the end, the framework of the improved version of the RO (IRO), its pseudo code, and the related flowchart are presented.

9.2.1 Snell's Light Refraction Law

Based on Snell's light refraction law, when light travels from a lighter medium to a darker one, it refracts, and its direction changes. Considering n_d and n_t as the refraction indexes of the lighter material and darker one, respectively, Snell's law can be expressed as:

$$n_d \cdot \sin\theta = n_t \cdot \sin\varnothing \tag{9.1}$$

where θ and Φ are the angles between the normal vector of the interface of two mediums (n) with incident ray vector (d) and the refracted ray vector (t), respectively, as shown in Fig. 9.1a.

Having the direction of the incident ray vector and the indexes of the refraction of the light and darker mediums, the direction of the refracted ray vector (t) can be found. The geometric proof for calculating t is easily accessible in the literature.

In a two-dimensional space, t can be calculated as the following in which, t, d, and n are unit vectors.

$$t = -n \cdot \sqrt{1 - \frac{n_d^2}{n_t^2} \cdot sin^2(\theta) + \frac{n_d}{n_t} \cdot (d - (d \cdot n) \cdot n)} \tag{9.2}$$

In a three-dimensional space, d and n are stated in a new coordinate system as:

$$n^* = (1, 0), d^* = (d \cdot i^*, d \cdot j^*) \tag{9.3}$$

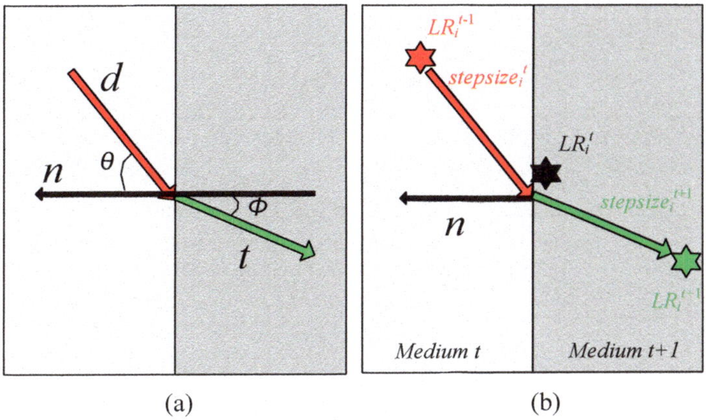

Fig. 9.1 Snell's light refraction law as the essence of the RO: (**a**) incident and refracted rays and their specifications and (**b**) transition of an agent in two consecutive iterations of the RO

Table 9.1 The components of the new coordinate system

	$-0.05 \leq n \cdot d \leq 0.05$	$0.05 \leq n \cdot d \leq 1$	$-1 \leq n \cdot d \leq -0.05$
i^*	n	n	n
j^*	d	$j^* = \frac{\left(n - \frac{d}{n \cdot d}\right)}{norm\left(n - \frac{d}{n \cdot d}\right)}$	$j^* = \frac{\left(n + \frac{d}{-n \cdot d}\right)}{norm\left(n + \frac{d}{-n \cdot d}\right)}$

where i^* and j^* are two normalized vectors from Table 9.1, in which *norm* is a function that provides the length of a vector. Then $t^* = (t_1^*, t_2^*)$ is calculated in two-dimensional space; now t can be obtained in a three-dimensional space:

$$t = t_1^* \cdot i^* + t_2^* \cdot j^* \tag{9.4}$$

9.2.2 *The Analogy Between Snell's Light Refraction Law and a Metaheuristic*

Consider a light ray which is crossing the transparent medium t, $stepsize_i^t$. It starts from the point LR_i^{t-1} and ends at the point LR_i^t. After refracting, it enters to the darker medium $t+1$ and continues its path with the vector $stepsize_i^{t+1}$ to reach a new point as LR_i^{t+1}. The refraction process is depicted in Fig. 9.2b. This new direction can be calculated for the two-dimensional and three-dimensional spaces based on Eqs. (9.2) and (9.4), respectively. They are presented as Snell's light refraction law at the previous subsection. Snell's light refraction in addition to the available information just needs the normal vector of the interface surface between two mediums (n) and also the ratio of their refraction indexes (n_d/n_t).

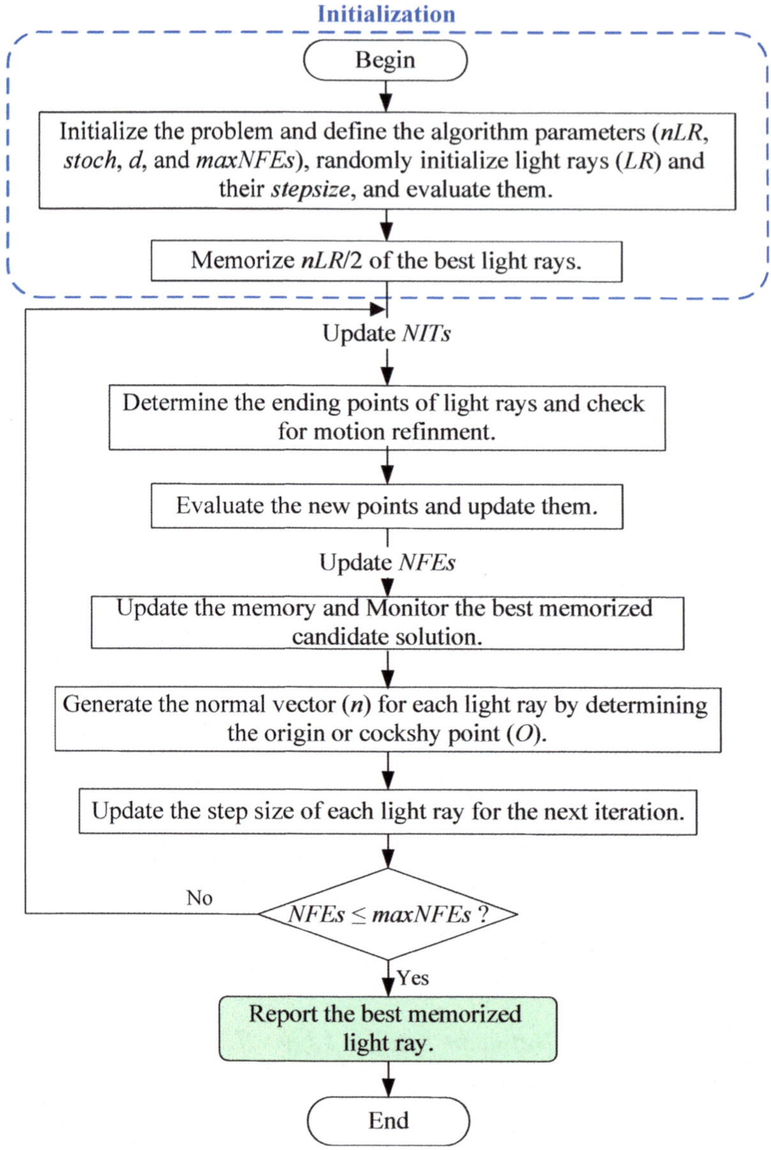

Fig. 9.2 Flowchart of the IRO algorithm

Looking at Fig. 9.1b, this schematic of Snell's law arises two sequential iterations of a metaheuristic in the mind as benefited by Kaveh and Khayatazad [1] to develop the RO. They select a value equal to 0.45 for the refraction index ratio. On the other hand, the vector n is considered so that it passes through two points. The ending point is the current position of the agent (LR_i^t), and the beginning point which is named as origin or Cockshy point, O_i^t, is defined as:

$$O_i^t = \frac{(maxNITs + NITs) \cdot BEST\_LR + (maxNITs - NITs) \cdot best\_LR_i}{2maxNITs} \quad (9.5)$$

where *maxNITs* is the total number of algorithm iterations which can be calculated by dividing the *maxNFEs* as the stopping criteria with the number of light rays and *BEST_LR* and $best\_LR_i$ are the best-so-far position found by all agents and the best position of *i*th agent found by itself, respectively. Such a definition, considering the normal vector so that it starts from a point determined based on the individual and collective knowledge of agents and ends in the current position of agents, creates a good balance between exploration and exploitation.

There are some other concepts in the framework of RO to guaranty the stochastic nature of the algorithm and also to return the violated agents into feasible search space, which will be defined at the next subsection. However, it should be noted that Snell' law is applicable to two- and three-dimensional search spaces. If the number of variables is more than 3, for using the ray tracing concept, the search space should be divided to a number of 2D and/or 3D spaces. In general, if nV is an even number, the search space is divided to $nV/2$ two-dimensional spaces, and if nV is an odd number, the search space is divided to $(nV - 3)/2$ two-dimensional space(s) and one 3D space. These subspaces should be merged to make a candidate solution and then evaluated. With this description, the step size vectors also should be divided to the corresponding 2D and or 3D spaces. The steps of Ray Optimization algorithm are presented at the following subsection.

9.2.3 Ray Optimization (RO) Algorithm

The steps of Ray Optimization algorithm are as follows.

Step 1: Initialization
Randomly initialize number of light rays (*nLR*): their beginning point matrix ($LR^{t=0}$) and step size matrix ($stepsize^{t=0}$) as:

$$LR_i^{t=0} = Lb + rand \times (Ub - Lb), i = 1, 2, \ldots, nLR \quad (9.6)$$

$$stepsize_i^{t=0} = -1 + 2 \times rand, i = 1, 2, \ldots, nLR \quad (9.7)$$

In the initialization phase, simply by adding the step size matrix to the beginning point matrix, the ending points of the initial light rays are obtained. These points will be the beginning points for the first iteration.

If an agent violates a boundary, it intersects the boundary at a specified point, because of having definite movement vector. Now using this point and the initial position of the agent, one can make a new vector whose direction is the same as the prior step size vector and its length is a multiple of the distance which is between the initial position and the boundary intersection. This multiple should naturally be less than 1. Since the best answer, especially in engineering problems, is close to the boundaries, it is considered as 0.9. Therefore, if an agent existed from the search space, the length of its light ray ($stepsize_i$) should be refined. The new length of light ray is equal to 0.9 times the distance of current agent position and the intersection with the boundary. This process is named as motion refinement. After motion refinement, or returning the existed agents inside the search space, evaluate them based on the objective function, and determine the best-so-far position found by all agents (*BEST_LR*) and the best position of *i*th agent found by itself so far ($best\_LR_i$). Go to the next step which is the cyclic body of the algorithm.

Step 2: Origin or Cockshy Point Making and Convergent Step

If the number of design variables (nV) is more than 3, the starting points (LR_i) and step size vector ($stepsize_i$) of each agent should be divided to k subgroups. If nV is an even number, the search space is divided to $nV/2$ two-dimensional spaces, and if nV is an odd number, the search space is divided to $(nV - 3)/2$ two-dimensional space(s) and one 3D space. For the *k*th subgroup in the *t*th iteration of the algorithm, these subgroups are shown as $LR_i^{(k,\ t)}$ and $stepsize_i^{(k,\ t)}$. Similarly, the *BEST_LR* and $best\_LR_i$ should be divided into k subgroups.

Determine the origin or Cockshy point for the *k*th subgroup of the *i*th agent ($O_i^{k,\ t}$) using Eq. (9.5). Determine the normal vector (n) for each *k*th subgroup starting from the origin and ending in the current position of the agent. Having the ratio of the refraction indexes (n_d/n_t), the vector n, and the previous step size vector, determine the new step size of refraction based on Snell's refraction law for 2D and 3D subgroups using Eqs. (9.2) and (9.4), respectively. But it is necessary to notice that n and the previous step size vector might not be unit vectors. These vectors must be normalized to use in Snell' refraction law. Now the new direction of the step size of movement for the *k*th subgroup of the *i*th agent in the next iteration ($t + 1$) is determined and is named as $stepsize_i^{(k,t+1)'}$. Considering three following subjects, some modifications should be made on this step size vector and the final form of this vector presented at the following equation.

1. Based on Snell's refraction law, it is a normalized vector, and it requires a logical coefficient.
2. In some cases, it is possible that, for an agent, the origin and its current position are the same, so the direction of normal cannot be obtained. This problem occurs when the agent is the best-so-far one. Therefore, it is logical to permit it to move in the same direction because of finding a more adequate answer, but the length of this vector should be changed.
3. One of the important features of each metaheuristic algorithm is that it should have a stochastic nature to find the best answer. RO considered this feature by adding a random change to the step size vector. In another word, there is a possibility like *stoch* that specifies whether it must be changed or not.

$$
RL_i^{(k,t)} \neq O_i^{(k,t)} \rightarrow \begin{cases} \text{w.p.}(1-stoch) \begin{cases} stepsize_i^{(k,t+1)} = stepsize_i^{(k,t+1)\prime} \times norm\left(RL_i^{(k,t)} - O_i^{(k,t)}\right) \\ stepsize_i^{(k,t+1)\prime} \text{ is determined by ray tracing.} \end{cases} \\ \text{w.p.} stoch \quad \begin{cases} stepsize_i^{(k,t+1)} = \dfrac{stepsize_i^{(k,t+1)\prime}}{norm\left(stepsize_i^{(k,t+1)\prime}\right)} \times \dfrac{a}{d} \times rand \\ stepsize_i^{(k,t+1)\prime} = -1 + 2 \times rand \end{cases} \end{cases}
$$

$$
RL_i^{(k,t)} = O_i^{(k,t)} \longrightarrow \quad stepsize_i^{(k,t+1)} = \frac{stepsize_i^{(k,t)}}{norm\left(stepsize_i^{(k,t)}\right)} \times rand \times 0.001. \tag{9.8}
$$

In this equation, *stoch* and *d* are 0.35 and 7.5, respectively, and *a* is calculated by:

$$
a = \sqrt{\sum_{i=1}^{n} (Lb_i - Ub_i)^2} \tag{9.9}
$$

where Lb_i and Ub_i are the maximum and minimum limits of variables that belong to the *i*th component of the step size vector and *n* is equal to 2 and 3 for 2D and 3D subgroups, respectively.

Merge the 2D and 3D $LR_i^{(k,\,t)}$ and $stepsize_i^{(k,\,t)}$ based on the subgrouping scheme, and form the agent matrix and its corresponding step size matrix. Update the agents using this step size matrix. Evaluate them and determine the *BEST_LR* and $best\_LR_i$.

Step 3: Stopping Criteria
If the stopping criterion of the algorithm is fulfilled, the search procedure terminates; otherwise the algorithm returns to the second step and continues the search. The finishing criterion is considered here as the maximum number of objective function evaluations (*maxNFEs*) or the maximum number of algorithm iterations (*maxNITs*).

9.2.4 Improved Ray Optimization (IRO) Algorithm

Kaveh et al. [6] developed an Improved Ray Optimization (IRO) algorithm employing a new approach in generating new solution vectors which has no limitation on the number of variables, so in the process of the algorithm, there is no need to divide the variables into groups like RO. The procedure which returns the violated agents into feasible search space is also modified. These improvements enhance the accuracy and convergence rate of RO. Here IRO is coded because of its efficiency and simplicity over RO algorithm.

At the following, firstly, the enhancements are defined, and then the steps of IRO are presented.

Enhancements:

1. The formulation of the origin is enhanced in a way that a memory is considered which saves the best position found by some or all the light rays and named as *LR_M*. If the number of light rays is more than or equal to 25, the size of memory is considered as 25; otherwise, it is taken as half number of the agents. The best position found by all agents so far (*bestLR*) will be the first array that is saved in the memory. Instead of using the best position visited by each light ray, a memorized position ($LR\_M^{(randi)}$) will be selected randomly and used in the formulation of origin as:

$$O_i^t = \frac{(maxNITs + NITs) \cdot bestLR + (maxNITs - NITs) \cdot LR\_M^{(randi)}}{2maxNITs} \tag{9.10}$$

2. In the RO if an agent existed from the search space, the length of its light ray ($stepsize_i$) should be modified which is named as motion refinement. The new length of light ray is equal to 0.9 times the distance of current agent position and the intersection with the boundary. In IRO instead of changing all the components of the violating agents, only the components that violate the boundary is refunded. This improvement is more effective when the number of variables is large. Motion refinement is formulated as:

$$LR_{ij}^{refined,\quad t+1} = \begin{cases} LR_{ij}^t + 0.9\left(Lb_j - LR_{ij}^t\right) & if\ \left(LR_{ij}^t < Lb_j\right) \\ LR_{ij}^t + 0.9\left(LR_{ij}^t - Ub_j\right) & if\ \left(LR_{ij}^t > Ub_j\right) \end{cases} \tag{9.11}$$

3. RO in addition to the available information (the previous step size vector) just needs the normal vector of the interface surface between two mediums (n) and also the ratio of their refraction indexes (n_d/n_t) to refract each light ray based on Snell's law. To achieve this aim, if the number of variables was more than 3, Snell's formula cannot be applied directly, and first, the main problem must be divided into some subproblems, and after the calculation, results of the subproblems should be merged to evaluate the goal function. When the number of variables is large, the computational cost grows considerably. Instead of this approach, the following formula (which has no limit on the number of variables) is utilized in IRO to calculate the direction of the new $stepsize'$ vector that is defined according to the prior *stepsize* vector and the normal vector:

$$stepsize_i^{(t+1)'} = \alpha \cdot n + \beta \cdot stepsize_i^{(t)} \tag{9.12}$$

where α and β are the factors that control the exploitation and exploration, respectively. An efficient optimization algorithm should perform good exploration in early iterations and good exploitation in the final iterations. Thus, α and β are increasing and decreasing functions, respectively, and are defined as:

$$\alpha = 1 + \left(\frac{NITs}{maxNITs}\right); \ \beta = 1 - 0.5\left(\frac{NITs}{maxNITs}\right) \tag{9.13}$$

As it is clear, this new formulation not only is simpler, but also Snell's law refracts the light vector gradually toward the normal vector. It should be noted that this formulation has only two parameters (α and β) same as the RO which needs the ratio of refraction indexes of two mediums (n_d/n_t).

Considering these enhancements, the steps of IRO can be presented in the same framework as the RO.

The Steps of Improved Ray Optimization Algorithm

Step 1: Initialization

After defining the optimization problem and the parameters of the algorithm, number of light rays (*nLR*): their beginning point matrix ($LR^{t=0}$) and step size matrix ($stepsize^{t=0}$) randomly initialized same as the RO using Eqs. (9.6) and (9.7). Evaluate the initial population, and form its corresponding vectors of the objective function (*Fit*) and the penalized objective function (*PFit*). It should be noted that the design vectors all are inside the search space. Form the light ray memory matrix (*LR_M*) and its corresponding vectors of objective function memory (*Fit_M*) and penalized objective function memory (*PFit_M*). It should be noted that in the initialization phase the memorizing process simply can be handled. Go to the next step which is the cyclic body of the algorithm.

Step 2: Origin or Cockshy Point Making and Convergent Step

Refract the matrix of light rays by adding the *stepsize* matrix to the beginning points. If each light ray existed from the search space, return it inside the search space using Eq. (9.11). Evaluate and update the new light rays, and update their corresponding vectors of the objective function (*Fit*) and penalized objective function (*PFit*). Update the light ray memory matrix (*LR_M*) and its corresponding vectors of objective function memory (*Fit_M*) and penalized objective function memory (*PFit_M*), if the light rays found and visit a better place in the search space. Using Eq. (9.10) determine the origin or Cauchy point for each light ray. It should be noted that the best position visited by all the light rays is the first array memorized in the *LR_M* matrix. Consequently, generate their corresponding normal vectors (n). The normal vectors start from the origin point and end in the current position of the light rays ($n_i = O_i - LR_i$). Using Eqs. (9.12) and (9.13), generate the *stepsize* for each light ray, and modify it based on Eq. (9.8) which will be used in the next iteration of the algorithm. It should be noted again that in the IRO dividing the design variables to k subgroups is not needed and k is equal to 1 here for any values of nV.

Step 3: Stopping Criteria

This step is same as the RO.

The pseudo code of IRO is provided as follows, and the flowchart is illustrated in Fig. 9.2.

The pseudo code of the IRO algorithm for solving COPs:

Define the algorithm parameters: the number of light rays (*nLR*), the probability of changing the refracted light ray vector (*stoch*), the *d* parameter which is used in the *stepsize* modification formula and should be considered equal to 7.5, and the maximum number of objective function evaluations as the stopping criteria (*maxNFEs*). Parameter *a* which is used in the *stepsize* modification formula can be easily determined by having the lower and upper bounds of design variables.
Generate random initial solutions or light rays: their begging points (*LR*) and the length and direction or the *stepsize*.
Evaluate initial solutions (*LR*); form its corresponding vectors of the objective function (*Fit*) and the penalized objective function (*PFit*).
Form the light ray memory matrix (*LR_M*) and its corresponding vectors of objective function memory (*Fit_M*) and penalized objective function memory (*PFit_M*).
While *NFEs*<*maxNFEs*

Update the number of algorithm iterations (*NITs*).
Determine the ending points of light rays.
Motion refinement.
Evaluate the new points and update the *LR* matrix and corresponding *Fit* and *PFit* vectors.
Update *NFEs*.
Update the *LR_M* matrix and corresponding *Fit_M* and *PFit_M* vectors.
Monitor the best memorized candidate solution (*bestLR*) and its corresponding objective (*minFit*) and penalized (*minPFit*) objective function.
Generate the normal vector (*n*) for each light ray by determining the origin or cockshy point (*O*).
Update the step size of each light ray for the next iteration.

end While

9.3 MATLAB Code for the Improved Ray Optimization (IRO) Algorithm

According to the Sect. 9.2.4, five functions are considered for coding the algorithm. First, these functions are presented with the order of their recall in the algorithm, and at the end, the algorithm is coded. The required comments are presented. Note that the percent sign (%) represents a comment.

The first function is the objective function evaluation (*fobj*) which was presented in Chap. 2.

The second function named as the *Motion_Refinment* corrects the position of ending points of the refracted light rays (*newLR*) if it swerves the side limits (*Lb* and *Ub*). The input arguments are starting points of the light rays (*LR*), ending points (*newLR*), and lower (*Lb*) and upper (*Ub*) bounds of design variables. The only output of the function is the refined ending point matrix. The *Motion_Refinment* function is coded as it follows. Notice that the existing dimensions will be checked again in the *fobj* function which is not needed. However, to have the same code of *fobj* used in the previous chapters, it remains unchanged.

```
% If a solution swerves the side limits correct its position using the
Motion_Refinment function.
function [newLR]=Motion_Refinment(LR,newLR,Lb,Ub)

for i=1:size(newLR,1)
    for j=1:size(newLR,2)
        if (newLR(i,j)<Lb(j))
            newLR(i,j)=LR(i,j)+0.9*(Lb(j)-LR(i,j));
        end
        if (newLR(i,j)>Ub(j))
            newLR(i,j)=LR(i,j)+0.9*(LR(i,j)-Ub(j));
        end
    end
end
```

The third function is the *Memory* function in which the memory matrix of light rays (*LR_M*) and its corresponding vectors of objective function memory (*Fit_M*) and penalized objective function memory (*PFit_M*) are updated. It should be noted that in the initialization phase the memorizing should take place firstly. However, the *Memory* function will not be called for memorizing in the initialization phase because of its simplicity in this phase.

```
% Update the light rays memory matrix (LR_M) and its corresponding vectors
of objective function memory (Fit_M) and penalized objective function
memory (PFit_M).
function [LR_M,Fit_M,PFit_M]=Memory(LR,Fit,PFit,LR_M,Fit_M,PFit_M)
nLR=size(LR,1);

for i=1:nLR
    for j=1:ceil(nLR/2)
        if PFit(i)<PFit_M(j)
            LR_M(j,:)=LR(i,:);
            Fit_M(j)=Fit(i);
            PFit_M(j)=PFit(i);
            break
        end
    end
end
```

The fourth function is determining the origin or Cauchy point (*O*) of each light ray and generating the corresponding normal vector (*n*) which is named as *Cockshy*. Input arguments are the current (*LR*) and memorized (*LR_M*) set of points, the best position visited by all the light rays until yet (*bestLR*), and the current (*NITs*) and the maximum (*maxNITs*) number of algorithm iterations. The output argument is the normal vector matrix of the light rays. The *Cockshy* function is coded as follows:

```
% Generate the origin and the normal vector using the Cockshy function.
function [n]=Cockshy(LR,bestLR,LR_M,NITs,maxNITs)

nLR=size(LR,1);

index=randi(size(LR_M,1),1,nLR);
for i=1:nLR
        O(i,:)=(maxNITs+NITs)/(2*maxNITs)*bestLR+(maxNITs-
NITs)/(2*maxNITs)*LR_M(index(i),:);
end

n=O-LR;
```

The fifth function is the *Snell* function by which the *stepsize* matrix is generated and modified to use in the next iteration for refracting the current set of light rays. The input arguments are the current set of *stepsize* matrix, the matrix of normal vectors (*n*), and the parameters needed in the Snell-based formulation (*a*, *d*, *stoch*, *NITs*, and *maxNITs*). The *Snell* function is coded as follows:

```
% Update the step size of each light ray for the next iteration using a
procedure like the Snell's rule.
function [stepsize]=Snell(stepsize,n,a,d,stoch,NITs,maxNITs)

nLR=size(stepsize,1);
nV=size(stepsize,2);

for i=1:nLR
    if norm(n(i,:))==0
        continue
    end
    n(i,:)=n(i,:)/norm(n(i,:));
    stepsize(i,:)=stepsize(i,:)/norm(stepsize(i,:));
end

alfa=1+(NITs/maxNITs);
beta=1-0.5*(NITs/maxNITs);

for i=1:nLR
    R=rand;
    if R<stoch
        stepsize(i,:)=-1+(2*rand(1,nV));
        stepsize(i,:)=(stepsize(i,:)/norm(stepsize(i,:)))*(a/d)*rand;
    end
    if R>=stoch
        if norm(n(i,:))<1e-7
            stepsize(i,:)=stepsize(i,:)*rand*0.001;
        else
            stepsize(i,:)=(alfa*n(i,:))+(beta*stepsize(i,:));
            stepsize(i,:)=stepsize(i,:)/norm(stepsize(i,:));
            stepsize(i,:)=stepsize(i,:)*(a/d);
        end
    end
end
```

The IRO algorithm is coded in the following and is composed of three parts: initialization, algorithm cyclic body, and monitoring the results.

```
clc
clear
% rng('default');
%% Initialization

% Define the properties of COP (tension/compression spring design problem).
nV=3; % Number of design variables.
Lb=[0.05 0.25 2]; % Lower bounds of design variables.
Ub=[2 1.3 15]; % Upper bounds of design variables.

% Define the parameters of IRO algorithm.
nLR=20; % Number of Light rays.
stoch=0.35;
d=7.5;
maxNFEs=40000; % Maximum Number of Objective Function Evaluations.
a=sqrt(sum((Lb-Ub).^2));

% Randomly generate nLR number of light rays as the initial population of
the algorithm.
for i=1:nLR
    LR(i,:)=Lb+(Ub-Lb).*rand(1,nV);  %Initial  beginning  points  of  light
rays.
    stepsize(i,:)=-1+2*rand(1,nV);  %  Initial  stepsize  (direction  and
length) of light rays.
end

% Evaluate initial population (LR) calling the fobj function constructed in
the second chapter and form its corresponding vectors of objective function
(Fit) and penalized objective function (PFit). It should be noted that the
design vectors all are inside the search space.
for i=1:nLR
    [X,fit,pfit]=fobj(LR(i,:),Lb,Ub);
    LR(i,:)=X;
    Fit(i)=fit;
    PFit(i)=pfit;
end

% Form the light rays memory matrix (LR_M) and its corresponding vectors of
objective function memory (Fit_M) and penalized objective function memory
(PFit_M). It should be noted that in the initialization phase it is not
needed to utilize the Memory function and memorizing simply can be done.
[value,index]=sort(PFit);
LR_M=LR(index(1:ceil(nLR/2)),:);Fit_M=Fit(index(1:ceil(nLR/2)));PFit_M=PFit
(index(1:ceil(nLR/2)));

%% Algorithm Body

NFEs=0;  %  Current  Number  of  Objective  Function  Evaluations  used  by  the
algorithm until yet.
NITs=0; % Number of algorithm iterations
maxNITs=maxNFEs/nLR; % Maximum Number of algorithm iterations

while NFEs<maxNFEs
    NITs=NITs+1; %Update the number of algorithm iterations.

    % Determine the ending points of light rays.
    newLR=LR+stepsize;

    % Motion refinement.
    [newLR]=Motion_Refinment(LR,newLR,Lb,Ub);

    % Evaluate the ending points of light rays calling the fobj function.
```

```
    % It should be noted that the existed dimensions are corrected before.
However, to do not change the form of fobj function it is checked again
here which is not needed.
    for i=1:nLR
        [X,fit,pfit]=fobj(newLR(i,:),Lb,Ub);
        LR(i,:)=X;
        Fit(i)=fit;
        PFit(i)=pfit;
    end

    % Update the number of objective function evaluations used by the
algorithm until yet.
    NFEs=NFEs+nLR;

    % Update the light rays memory matrix (LR_M) and its corresponding
vectors of objective function memory (Fit_M) and penalized objective
function memory (PFit_M) utilizing the Memory function.
    [LR_M,Fit_M,PFit_M]=Memory(LR,Fit,PFit,LR_M,Fit_M,PFit_M);

    % Monitor the best candidate solution (bestLR) found by the algorithm
until yet, and its corresponding objective function (minFit) and penalized
objective funtion (minPFit).
    bestLR=LR_M(1,:); minFit=Fit_M(1); minPFit=PFit_M(1);

    % Generate the normal vector (n) for each light ray by determining the
origin or cockshy point (O).
    [n]=Cockshy(LR,bestLR,LR_M,NITs,maxNITs);

    % Update the step size of each light ray for the next iteration.
    [stepsize]=Snell(stepsize,n,a,d,stoch,NITs,maxNITs);

    % Display desired information of the iteration.
    disp(['NITs= ' num2str(NITs) '; minFit = ' num2str(minFit) '; minPFit
= ' num2str(minPFit)]);

    % Save the required results for post processing and visualization of
the algorithm performance.
    output1(NITs,:)=[minFit,minPFit,NFEs];
    output2(NITs,:)=[min(PFit),max(PFit),mean(PFit)];
    output3(NITs,:)=[bestLR,NFEs];
end

%% Monitoring the results
figure;
plot((1:1:NITs),output2(:,1),'g',(1:1:NITs),output2(:,2),'r--
',(1:1:NITs),output2(:,3),'b-.')
legend('min','max','mean');
xlabel('NITs');
ylabel('PFit');
```

9.4 Experimental Evaluation

IRO needs four parameters to be tuned: the number of light rays (*nLR*), the probability of changing the refracted light ray vector (*stoch*) recommended equally to 0.35, the *d* parameter which is used in the *stepsize* modification formula and recommended as 7.5, and the maximum number of objective function evaluations as the stopping criteria (*maxNFEs*). Another parameter defined as a which is used in the *stepsize* modification formula can be easily determined by having the lower and upper bound of design variables. It should be also noted that value equal to 0.9 in the formulation of motion refinement can be considered also as another parameter. There is also another parameter in the original RO as the ratio of the refraction indexes of the two mediums which is considered as two parameters (α and β) in IRO. α and β are defined adaptively in the IRO.

Considering a large enough value (20,000) for *maxNFEs* to ensure convergence of the algorithms and making the sensitivity analysis, it is found that the recommended values in the original papers are almost efficient, and considering the *stoch* parameter is inevitable to ensure the diversification of the algorithm, the *nLR* between 15 and 40 results in a good performance (accuracy and convergence speed), and also the defined formulas for the values of α and β are efficient in balancing the diversification and intensification ability of the algorithm.

References

1. Kaveh A, Khayatazad M (2012) A new meta-heuristic method: ray optimization. Comput Struct 112:283–294
2. Kaveh A, Khayatazad M (2013) Ray optimization for size and shape optimization of truss structures. Comput Struct 117:82–94
3. Kaveh A, Javadi SM (2014) Shape and size optimization of trusses with multiple frequency constraints using harmony search and ray optimizer for enhancing the particle swarm optimization algorithm. Acta Mech 225:1595–1605
4. Kaveh A, Javadi SM (2014) An efficient hybrid particle swarm strategy, ray optimizer, and harmony search algorithm for optimal design of truss structures. Period Polytech Civ Eng 58:155
5. Kaveh A, Ilchi Ghazaan M (2015) Layout and size optimization of trusses with natural frequency constraints using improved ray optimization algorithm. Iran J Sci Technol Trans Civ Eng 39:395–408
6. Kaveh A, Ilchi Ghazaan M, Bakhshpoori T (2013) An improved ray optimization algorithm for design of truss structures. Period Polytech Civ Eng 57(2):97–112

Chapter 10
Colliding Bodies Optimization Algorithm

10.1 Introduction

Colliding Bodies Optimization (CBO) algorithm is a new metaheuristic search algorithm developed by Kaveh and Mahdavi [1]. CBO is inspired based on the physics laws of momentum and energy that govern the collisions accrued between solid bodies. CBO is simple in concept and does not depend on any internal parameters. In this regard, it has been used to solve a wide variety of optimization problems encountered in practice. CBO also has been used extensively in the field of structural optimization problems [2–8].

The essence of CBO is that of combining simple operations taken from the physics laws of momentum and energy that govern the collisions accrued between solid bodies as to solve complex problems. The key is that one object collides with another object and they move toward a minimum energy level. Like other metaheuristics, CBO starts from a set of randomly generated initial candidate solutions. In each iteration of the algorithm, all population is evaluated and sorted according to their objective function. The first half of the best are considered as objects with the zero velocity as the stationary objects. The remaining half are considered as moving objects with a specific velocity. The moving objects will be collided to the stationary ones with the aim of reaching themselves to the better position and also pushing the stationary ones to the better positions in the search space. The new positions of the objects are attained simply using the new velocities which are computable from physics laws of momentum and energy that govern the collisions accrued between solid bodies.

A. Kaveh, T. Bakhshpoori, *Metaheuristics: Outlines, MATLAB Codes and Examples*, https://doi.org/10.1007/978-3-030-04067-3_10

10.2 Formulation and Framework of the CBO

In the following, firstly, the physics laws of momentum and energy that govern the collisions accrued between solid bodies as the essence of CBO are presented. Then the concepts and framework of the CBO algorithm, its pseudo code, and the related flowchart are presented.

10.2.1 The Collision Between Two Bodies

Conservation laws in physics for the collision of two bodies with the masses of m_1 and m_2, the initial velocities of v_1 and v_2, the secondary velocities of v'_1 and v'_2, and the Q as the loss of kinetic energy due to the impact are as follows:

$$m_1 \cdot v_1 + m_2 \cdot v_2 = m_1 \cdot v'_1 + m_2 \cdot v'_2 \tag{10.1}$$

$$\frac{1}{2}m_1 v_1^2 + \frac{1}{2}m_2 v_2^2 = \frac{1}{2}m_1 v_1'^2 + \frac{1}{2}m_2 v_2'^2 \tag{10.2}$$

The secondary velocities can be calculated as:

$$\begin{aligned} v'_1 &= \frac{(m_1 - \varepsilon m_2)v_1 + (m_2 + \varepsilon m_2)v_2}{m_1 + m_2} \\ v'_2 &= \frac{(m_2 - \varepsilon m_1)v_2 + (m_1 + \varepsilon m_1)v_1}{m_1 + m_2} \end{aligned} \tag{10.3}$$

where ε is the coefficient of restitution (COR) of two colliding bodies, defined as ratio of the relative velocity of separation to the relative velocity of approach:

$$\varepsilon = \frac{|v'_2 - v'_1|}{|v_2 - v_1|} = \frac{v'}{v} \tag{10.4}$$

According to the coefficient of restitution, two special cases of collision can be considered: (1) a perfectly elastic collision is defined as the one in which there is no loss of kinetic energy in the collision ($Q = 0$ and $\varepsilon = 1$). In reality, any macroscopic collision between objects will convert some kinetic energy to the internal and other forms of energy. In this case, after the collision, the velocity of separation is high. (2) An inelastic collision is the one in which part of the kinetic energy is changed to some other forms of energy in the collision. Momentum is conserved in inelastic collisions (as it is for elastic collisions), but one cannot track the kinetic energy through the collision since some of it is converted to other forms of energy. In this case, the coefficient of restitution does not equal to one ($Q \neq 0$ and $\varepsilon \leq 1$). Here, after the collision, the velocity of separation is low. For most of the real objects, ε is between 0 and 1.

10.2.2 The CBO Algorithm

Like other population-based metaheuristics, CBO starts from a set of candidate solutions randomly generated within the search space. The number of colliding bodies is considered as *nCB*. These objects form the matrix of colliding bodies (*CB*). After evaluating the objects, the corresponding objective function (*Fit*) and the penalized objective function (*PFit*) are produced. For minimization problems it is suitable to consider larger values of mass for the candidate solutions with the less penalized objective function. Therefore, the mass of each colliding body is calculated using the following formula and forms the mass matrix of the objects (*M*):

$$M(i) = \frac{1/PFit(i)}{\sum_{i=1}^{nCB} 1/PFit(i)} \tag{10.5}$$

For colliding, the objects should be categorized into two groups: the stationary objects and the moving objects. More efficient concept to categorize the bodies would be that the better ones are considered as the stationary objects. In this way, the objects are sorted in an ascending order based on their penalized objective function. The first half ($i = 1, \ldots, nCB/2$) are considered as the stationary objects, and their velocity will be zero before the collision. The remaining half ($i = nCB/2 + 1, \ldots, nCB$) will form the moving objects. These bodies will move toward to the corresponding stationary objects for collision. Therefore, their velocity before the collision is considered as:

$$v(i) = CB(i) - CB(i - nCB/2), \quad i = nCB/2 + 1, \ldots, nCB \tag{10.6}$$

Now having the masses of objects and their velocity before the collision, their new velocity after the collision can be simply calculated based on Eqs. (10.3) and (10.4). Let us consider the new velocities of objects after collision as the step size. The step size for each stationary object is as follows:

$$stepsize(i) = \frac{(M(i + nCB/2) + \varepsilon\, M(i + nCB/2))v(i + nCB/2)}{M(i) + M(i + nCB/2)}, \quad i = 1, \ldots, nCB/2 \tag{10.7}$$

and the step size for each moving object is as follows:

$$stepsize(i) = \frac{(M(i) - \varepsilon\, M(i - nCB/2))v(i)}{M(i) + M(i - nCB/2)}, \quad i = nCB/2 + 1, \ldots, nCB \tag{10.8}$$

As mentioned previously, ε is the coefficient of restitution (COR), and for most of the real objects, its value is between 0 and 1. It was defined as the ratio of the separation velocity of two agents after the collision to the approach velocity of two agents before the collision. In the CBO, this index is used to control the exploration and exploitation rate. For this goal, the COR decreases linearly from unit to zero. Thus, ε is defined as:

$$\varepsilon = 1 - \frac{NITs}{maxNITs} \tag{10.9}$$

where *NITs* is the current number of algorithm iterations and *maxNITs* is the maximum number of algorithm iterations considered as the stopping criteria of the algorithm.

CBO considers the current position of the stationary bodies as the origin of movement for both the stationary and moving objects. Therefore, the new position of both stationary and moving objects will be obtained by adding their new velocity or the step size to their current positions. However, to achieve a more dynamic search, a random part is also added to the formulation as the following:

$$newCB(i) = \begin{cases} CB(i) + rand \times stepsize(i) & i = 1, \ldots, nCB/2 \\ CB(i - nCB/2) + rand \times stepsize(i) & i = nCB/2 + 1, \ldots, nCB \end{cases} \tag{10.10}$$

The pseudo code of CBO is given as follows, and the flowchart is illustrated in Fig. 10.1.

The pseudo code of the CBO algorithm for solving COPs:

Define the algorithm parameters: *nCB* and *maxNFEs*.
Generate random initial colliding bodies (*CB*).
Evaluate the initial colliding bodies.
While *NFEs*<*maxNFEs*

Determine the mass matrix (*M*) of the objects using Eq. (10.5).
Determine the new velocity matrix after collision (*stepsize*) using Eqs. (10.7) and (10.8).
Update *NFEs*.

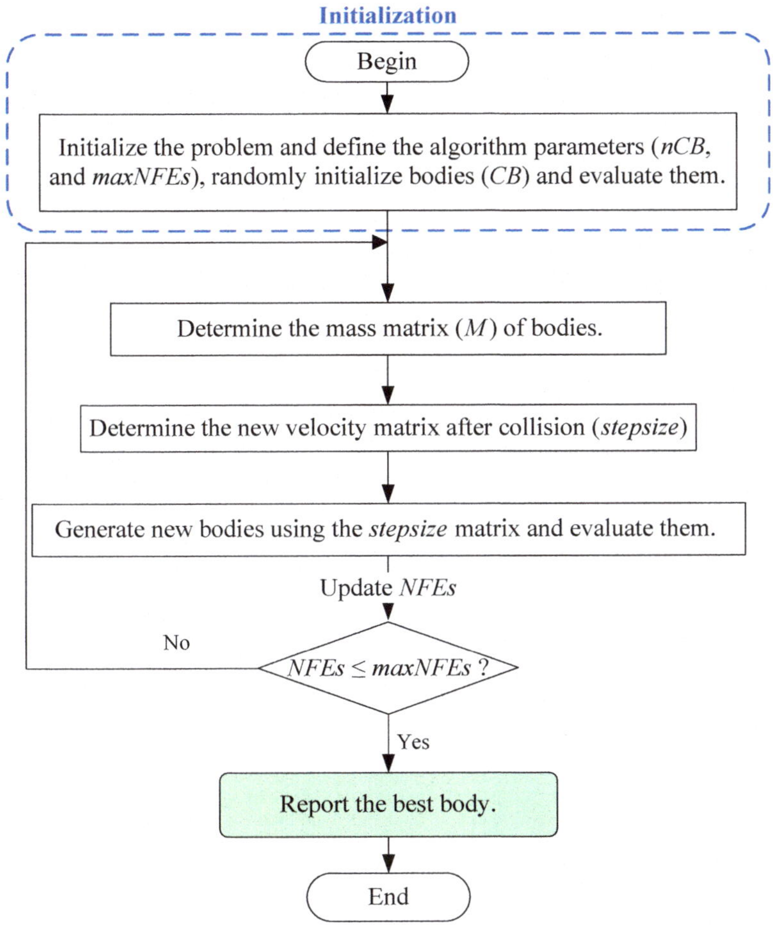

Fig. 10.1 Flowchart of the CBO algorithm

Upgrade bodies or generate new bodies using the *stepsize* matrix based on the Eq. (10.10).
Evaluate the new bodies. It should be noted that in the CBO algorithm the replacement strategy is not used.
Update *NFEs*.
Monitor best of the so far candidate solution.

end While

10.3 MATLAB Code for Colliding Bodies Optimization (CBO) Algorithm

As was mentioned before, the CBO has a simple structure. According to the previous section, two functions are considered for coding the algorithm. First, these two functions are presented with the order of their recall in the algorithm, and at the end, the algorithm is coded. The required comments are presented. Note that the percent sign (%) represents a comment.

The first function is the objective function evaluation (*fobj*) which was presented in Chap. 2.

The second function is named as the *Collision* functions to determine the new velocity of bodies after the collision. The input arguments are the current position of colliding bodies matrix (*CB*) and their corresponding objective (*Fit*) and penalized objective function vectors (*PFit*), the mass matrix (*M*), and the current (*NITs*) and the maximum (*maxNITs*) number of algorithm iterations. The input arguments are *CB*, *Fit*, *PFit*, *M*, and the new velocity or the *stepsize* matrix.

```
% Calculate the velocity of bodies after collision.
function [CB,M,Fit,PFit,stepsize]=Collision(CB,M,Fit,PFit,NITs,maxNITs)
nCB=size(CB,1);

[~,order]=sort(PFit);
CB=CB(order,:);
Fit=Fit(order);
PFit=PFit(order);
M=M(order);

COR=1-(NITs/maxNITs); % Coefficient of Restitution.
for i=1:nCB/2
    VMB=CB(nCB/2+i,:)-CB(i,:);   %   Velocity   of   moving   bodies   before
collision.
    stepsize(i,:)=((1+COR)*M(nCB/2+i))/(M(i)+M(nCB/2+i))*VMB; % Velocity of
stationary bodies after collision.
    stepsize(nCB/2+i,:)=(M(nCB/2+i)-COR*M(i))/(M(i)+M(nCB/2+i))*VMB;        %
Velocity of moving bodies after collision
end
```

The CBO algorithm is coded in the following and is composed of three parts: initialization, algorithm cyclic body, and monitoring the results.

```
clc
clear
%% Initialization

% Define the properties of COP (tension/compression spring design problem).
nV=3; % Number of design variables.
Lb=[0.05 0.25 2]; % Lower bounds of design variables.
Ub=[2 1.3 15]; % Upper bounds of design variables.

% Define the parameters of CBO algorithm.
nCB=20; % Number of colliding bodies.
maxNFEs=20000; % Maximum Number of Objective Function Evaluations.


% Randomly generate nCB number of colliding bodies as the initial
population of the algorithm.
for i=1:nCB
    CB(i,:)=Lb+(Ub-Lb).*rand(1,nV); % Initial beginning points of light
rays.
end

% Evaluate initial population (CB) calling the fobj function constructed in
the second chapter and form its corresponding vectors of objective function
(Fit) and penalized objective function (PFit). It should be noted that the
design vectors all are inside the search space.
for i=1:nCB
    [X,fit,pfit]=fobj(CB(i,:),Lb,Ub);
    CB(i,:)=X;
    Fit(i)=fit;
    PFit(i)=pfit;
end

% Monitor best the so far candidate solution (BestCB) and its corresponding
penalized objective function (MinPFit) and objective function (MinFit).
[MinPFit,m]=min(PFit);
MinFit=Fit(m);
BestCB=CB(m,:);

%% Algorithm Body

NFEs=0; % Current Number of Objective Function Evaluations used by the
algorithm until yet.
NITs=0; % Number of algorithm iterations
maxNITs=maxNFEs/nCB; % Maximum Number of algorithm iterations

while NFEs<maxNFEs
    NITs=NITs+1; %Update the number of algorithm iterations.

    % Form the mass matrix (M) of the objects.
    M=(1./PFit)./(sum(1./PFit));

    % Determine the new velocity matrix after collision.
    [CB,M,Fit,PFit,stepsize]=Collision(CB,M,Fit,PFit,NITs,maxNITs);

    for i=1:nCB
        if i<=nCB/2
            newCB(i,:)=CB(i,:)+rand(1,nV).*stepsize(i,:);
        else
            newCB(i,:)=CB(i-nCB/2,:)+rand(1,nV).*stepsize(i,:);
        end
    end
```

```
    % Evaluate the new bodies. It should be noted that in the CBO algorithm
the replacement strategy is not used.
    for i=1:nCB
        [X,fit,pfit]=fobj(newCB(i,:),Lb,Ub);
        CB(i,:)=X;
        Fit(i)=fit;
        PFit(i)=pfit;
    end

    % Update the number of Objective Function Evaluations used by the
algorithm until yet.
    NFEs=NFEs+nCB;

    %Monitor the best current candidate solution (bestCB) and its
corresponding penalized objective function (minPFit) and objective function
(minFit).
    [minPFit,m]=min(PFit);
    minFit=Fit(m);
    bestCB=CB(m,:);

    % Monitor best of the so far candidate solution (BestCB) and its
corresponding penalized objective function (MinPFit) and objective function
(MinFit).
    if minPFit<=MinPFit
        BestCB=bestCB;
        MinPFit=minPFit;
        MinFit=minFit;
    end

    % Display desired information of the iteration.
    disp(['NITs= ' num2str(NITs) '; MinFit = ' num2str(MinFit) '; MinPFit
= ' num2str(MinPFit)]);

    % Save the required results for post processing and visualization of
the algorithm performance.
    output1(NITs,:)=[MinFit,MinPFit,NFEs];
    output2(NITs,:)=[min(PFit),max(PFit),mean(PFit)];
    output3(NITs,:)=[BestCB,NFEs];
end

%% Monitoring the results
figure;
plot((1:1:NITs),output2(:,1),'g',(1:1:NITs),output2(:,2),'r--
',(1:1:NITs),output2(:,3),'b-.')
legend('min','max','mean');
xlabel('NITs');
ylabel('PFit');
```

10.4 Experimental Evaluation

As it was mentioned, the CBO uses no parameters except the ones needed for all population-based metaheuristics: the number of population (*nCB*) and the maximum number of objective function evaluations (*maxNFEs*) as the stopping criteria. Considering a large enough value for *maxNFEs* to ensure convergence of the algorithms is essential. This parameter is almost problem dependent and should be considered larger for complicated and larger problems. It is considered here equal to 20,000. The simulation results show that considering at least 10 numbers of bodies are requisite for the CBO, the number of bodies between 20 and 60 leads to acceptable performance of the algorithm and increasing the *nCB* requires higher computational cost without losing accuracy.

10.5 Extension to CBO

Enhanced Colliding Bodies Optimization has been developed to improve the performance of the CBO yet preserving its simple structure [9]. Colliding memory (CM) is added to save some historically best CB vectors and their related mass and objective function values to reduce the computational cost. In each iteration, the solution vectors saved in CM are added to the population, and the same numbers of current worst CBs are deleted. To prevent premature convergence, a parameter *Pro* within (0, 1) is introduced, and it is specified whether a component of each CB must be changed or not. For each colliding body, *Pro* is compared with *rand* which is a random number uniformly distributed within (0, 1). If $rand < Pro$, one dimension of the *i*th CB is selected randomly, and its value is regenerated. For further information the reader can refer to [9].

References

1. Kaveh A, Mahdavi VR (2014) Colliding bodies optimization: a novel meta-heuristic method. Comput Struct 139:18–27
2. Kaveh A, Mahdavi VR (2015) Colliding bodies optimization: extensions and applications. Springer, Cham
3. Kaveh A, Maniat M, Arab Naeini M (2016) Cost optimum design of post-tensioned concrete bridges using a modified colliding bodies optimization algorithm. Adv Eng Softw 98:12–22
4. Shayanfar MA et al (2016) Damage detection of bridge structures in time domain via enhanced colliding bodies optimization. Int J Optim Civ Eng 6:211–226
5. Kaveh A, Ardalani S (2016) Cost and CO2 emission optimization of reinforced concrete frames using enhanced colliding bodies algorithm. Asian J Civ Eng 17:831–858
6. Kaveh A, Moradveisi M (2016) Nonlinear analysis based optimal design of double-layer grids using enhanced colliding bodies optimization method. Struct Eng Mech 58:555–576
7. Kaveh A, BolandGerami A (2017) Optimal design of large-scale space steel frames using cascade enhanced colliding body optimization. Struct Multidiscipl Optim 55:237–256
8. Kaveh A, Sepehr S (2018) Structural optimization of jacket supporting structures for offshore wind turbines using colliding bodies optimization algorithm. Struct Des Tall Spec Build 27: e1494. https://doi.org/10.1002/tal.1494
9. Kaveh A, Ilchi Ghazaan M (2014) Enhanced colliding bodies optimization for design problems with continuous and discrete variables. Adv Eng Softw 77:66–75

Chapter 11
Tug of War Optimization Algorithm

11.1 Introduction

Kaveh and Zolghadr [1] developed a novel population-based metaheuristic algorithm inspired by the game tug of war. Utilizing a sports metaphor, the algorithm, denoted as tug of war optimization (TWO), considers each candidate solution as a team participating in a series of rope pulling competitions. The teams exert pulling forces on each other based on the quality of the solutions they represent. The competing teams move to their new positions according to Newtonian laws of mechanics. Unlike many other metaheuristic methods, the algorithm is formulated in such a way that considers the qualities of both of the interacting solutions. TWO is applicable to global optimization of discontinuous, multimodal, non-smooth, and non-convex functions. Viability of the TWO in solving engineering COPs currently is examined using some structural optimization problems [2–4]. It should be noted that TWO has a very simple structure and the available studies indicate its potential to be used for many real-size COPs.

Like other metaheuristics, TWO starts from a set of randomly generated initial candidate solutions. Each candidate solution is considered as a team, and all population form a league. In each iteration of the algorithm, after evaluation, teams are assigned by a value of weight proportional to their merit. Accordingly, the best team has the highest weight, and the worst one has the lowest weight. Consider two candidate solutions or two teams as two rival sides pulling a rope. As a rule, the lightest team will lose the competition and move toward the heaviest team. TWO idealizes the tug of war framework, to determine the resultant force affecting the defeated team due to its interaction with the heavier team. As a result, the lightest team accelerates toward the heaviest team according to Newton's second law with a calculable displacement rate based on the constant acceleration movement equation. TWO also added randomness to the formulation extracted from the Newtonian laws of mechanics. Analogously, in each iteration, the league is updated by a series of team-team rope pulling competitions with the aim of attracting teams toward the

A. Kaveh, T. Bakhshpoori, *Metaheuristics: Outlines, MATLAB Codes and Examples*, https://doi.org/10.1007/978-3-030-04067-3_11

optimum position. This forms the convergence operator of TWO. It should be noted that TWO additionally benefits a heuristic strategy to regenerate the teams are moved outside the search space which is most likely to be accrued for the lightest ones.

11.2 Formulation and Framework of TWO

In the following, after presenting the idealized tug of war framework as the essence of the TWO, the concepts and framework of TWO algorithm, its pseudo code, and the related flowchart are presented.

11.2.1 Idealized Tug of War Framework

Tug of war or rope pulling is a strength contest in which two competing teams pull on the opposite ends of a rope in an attempt to bring the rope in their direction against the pulling force of the opposing team. The activity dates back to ancient times and has continued to exist in different forms ever since. There has been a wide variety of rules and regulations for the game, but the essential part has remained almost unaltered. Naturally, as far as both teams sustain their grips of the rope, movement of the rope corresponds to the displacement of the losing team.

Triumph in a real game of tug of war generally depends on many factors and could be difficult to analyze. However, an idealized framework is utilized in TWO where two teams having weights W_i and W_j are considered as two objects lying on a smooth surface as shown in Fig. 11.1.

As a result of pulling the rope, the teams experience two equal and opposite forces (F_p) according to Newton's third law. For object i, as far as the pulling force is smaller than the maximum static friction force ($W_i\mu_s$), the object rests in its place. Otherwise the non-zero resultant force can be calculated as:

$$F_r = F_p - W_i \, \mu_k \tag{11.1}$$

As a result, the object i accelerates toward the object j according to Newton's second law:

$$a = F_r/(W_i/g) \tag{11.2}$$

Since the object i starts from a zero velocity, its new position can be determined as:

$$X_i^{new} = \frac{1}{2} a \, t^2 + X_i^{old} \tag{11.3}$$

Fig. 11.1 An idealized tug of war framework

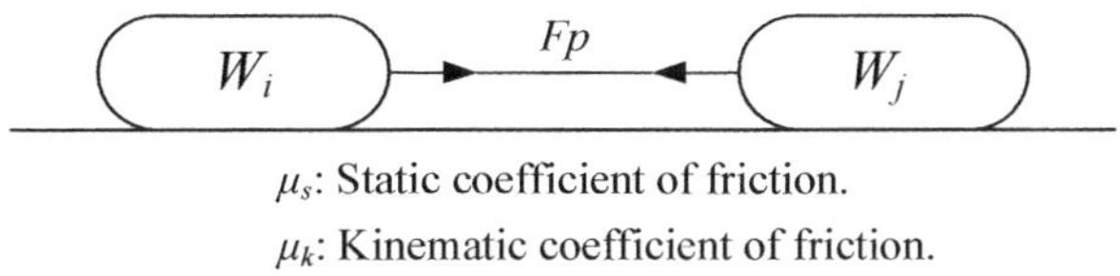

11.2.2 TWO Algorithm

Looking at the idealized tug of war framework overviewed at the previous subsection, there is an analogy between it and a population-based metaheuristic. Each agent of the algorithm can be considered as a team engaged in a series of tug of war competitions. The weight of teams is determined based on the quality of the corresponding solutions, and the amount of pulling force that a team can exert on the rope is assumed to be proportional to its weight. Naturally, the opposing team will have to maintain at least the same amount of force in order to sustain its grip on the rope. The lighter team accelerates toward the heavier team, and this forms the convergence operator of TWO. The algorithm improves the quality of the solutions iteratively by maintaining a proper exploration/exploitation balance using the described convergence operator. The rules of TWO can be stated as follows:

Rule 1: Initialization
TWO starts from a set of candidate solutions or teams randomly generated within the search space. The number of teams is considered as nT. These particles construct the matrix of teams (T) or the league matrix. After evaluating the teams, the corresponding objective function (Fit) and the penalized objective function ($PFit$) vectors are produced.

Rule 2: Weight Assignment
Each solution is considered as a team with the following weight:

$$W_i = \frac{PFit_i - \min(PFit)}{\min(PFit) - \max(PFit)} + 1 \tag{11.4}$$

where min and max return the minimum and the maximum element of the penalized objective vector, respectively. According to this definition, the weights of the teams range between 1 and 2 so that the best is the heaviest.

Rule 3: Competition and Displacement
In TWO each of the teams of the league competes against all the others one at a time to move to its new position in each iteration. The pulling force exerted by a team is assumed to be equal to its static friction force ($W\mu_s$). μ_s is the static coefficient of friction and can be considered as 1. Hence the pulling force between teams i and j ($F_{p,ij}$) can be determined as $\max\{W_i\mu_s, W_j\mu_s\}$. Such a definition keeps the position of the heavier team unaltered. Then the resultant force affecting team i due to its interaction with heavier team j can be calculated as follows:

$$F_{r,ij} = F_{p,ij} - W_i\, \mu_k \tag{11.5}$$

where μ_k is the kinematic coefficient of friction and can be considered equal to 1. Consequently, team i accelerates toward team j:

$$a_{ij} = \left(F_{r,ij}/W_i\mu_k\right)g_{ij} \tag{11.6}$$

where g_{ij} is the gravitational acceleration constant defined as:

$$g_{ij} = T_j - T_i \tag{11.7}$$

in which T_j and T_i are the position vectors for teams j and i. Finally, the displacement of the team i after competing with team j can be derived as:

$$stepsize_{ij} = \frac{1}{2}a_{ij}\Delta T^2 + \alpha\,\beta\,(Lb - Ub)\circ randn \tag{11.8}$$

The second term of this formulation introduces randomness into the algorithm. This term can be interpreted as the random portion of the search space traveled by team i before it stops after the applied force is removed. The role of α is to gradually decrease the random portion of the team's movement. For most of the applications, α could be considered as a constant chosen from the interval [0.9, 0.99]; bigger values of α decrease the convergence speed of the algorithm and help the candidate solutions explore the search space more thoroughly. β is a scaling factor which can be chosen from the interval (0,1). This parameter controls the steps of the candidate solutions when moving in the search space. When the search space is supposed to be searched more accurately with smaller steps, smaller values should be chosen for this parameter. *Lb* and *Ub* are the vectors containing the lower and upper bounds of the permissible ranges of the design variables, respectively; $\circ$ denotes element by element multiplication; *randn* is a vector of random numbers drawn from a standard normal distribution. ΔT is the time step and can be considered equal to 1.

It should be noted that when team j is lighter than team i, the corresponding displacement of team i will be equal to zero. Finally, the total displacement of team i is as follows (i not equal j):

$$stepsize_i = \sum_{j=1}^{nT} stepsize_{ij} \tag{11.9}$$

The new position of the team i is then calculated as:

$$T_i^{new} = T_i + stepsize_i \tag{11.10}$$

Rule 4: Updating the League

Once the teams of the league compete against each other for a complete round, the league should be updated. This is done by comparing the new candidate solutions (the new positions of the teams) with the current teams of the league. That is to say if the new candidate solution i is better than the worst team of the league in terms of penalized objective function value, the worst team is removed from the league, and the new solution takes its place.

Rule 5: Handling the Side Constraints

It is possible for the candidate solutions to leave the search space, and it is important to deal with such solutions properly. This is especial case for the solutions corresponding to lighter teams for which the values of *stepsize* are usually bigger. Different strategies might be used in order to solve this problem. For example, such candidate solutions can be simply brought back to their previous feasible position (flyback strategy), or they can be regenerated randomly. TWO uses a new strategy incorporating the best-so-far solution (*bestT*) to handle the dimensions of teams which exit side constraints. This strategy is utilized with a certain probability (0.5). For the rest of the cases, the violated limit is taken as the new value of the *j*th design variable. Based on this strategy, which is named as best team strategy (BTS), the new value of the *j*th design variable of the *i*th team that violated side constraints in the *NITs* iteration of the algorithm is defined as:

$$T_{ij} = bestT_j + (randn/NITs)\left(bestT_j - T_{ij}\right) \tag{11.11}$$

where *randn* is a random number drawn from a standard normal distribution. There is a very slight possibility for the newly generated variable to still be outside the search space. In such cases, a flyback strategy is used.

Rule 6: Termination Criteria

A maximum number of objective function evaluations (*maxNFEs*) or a maximum number of algorithm iterations (*maxNITs*) are considered as the terminating criterion.

Based on the six rules as the essence of the TWO, the pseudo code of TWO is given as follows, and the flowchart is illustrated in Fig. 11.2.

The pseudo code of TWO algorithm for solving COPs:

Define the algorithm parameters: nT, μ_s, μ_k, α, β, ΔT, and *maxNFEs*.
Generate random initial teams or the random initial league (T).
Evaluate the initial league and form its corresponding vectors of the objective function (*Fit*) and penalized objective function (*PFit*).
Monitor the best candidate solution or team (*bestT*) and its corresponding objective function (*minFit*) and penalized objective function (*minPFit*).
While *NFEs*<*maxNFEs*

Update the number of algorithm iterations (*NITs*).
Determine the weights of the teams of the league using Eq. (11.4).
Determine the total displacement of teams by a series of rope pulling competitions based on the equations presented in Rule 3.

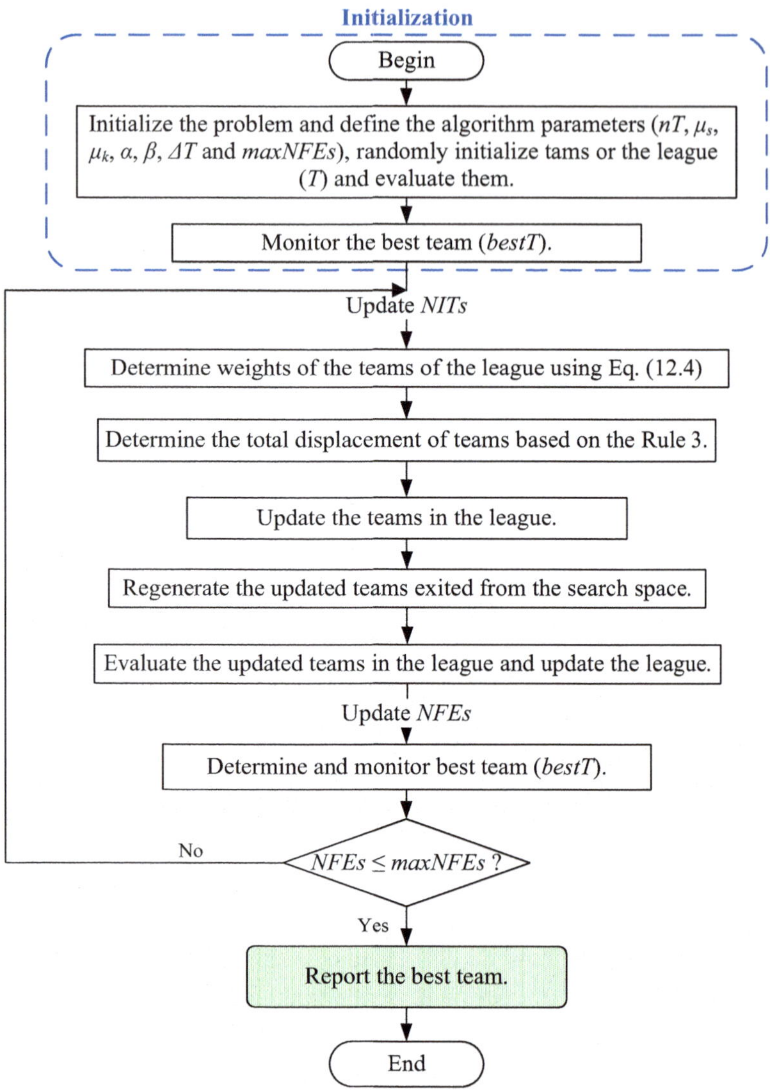

Fig. 11.2 Flowchart of the TWO algorithm

Update the teams in the league by adding the total displacement to their current position.

Regenerate the updated teams exited from the search space using the best team strategy.

Evaluate the updated teams in the league.

Update $NFEs$.

Update the league by replacing the current worst teams with the better teams produced.
Determine and monitor the best team (*bestT*) and its corresponding objective function (*minFit*) and penalized objective function (*minPFit*).

end While

11.3 Matlab Code for the Tug of War Optimization (TWO) Algorithm

According to the previous section, five functions are considered for coding the algorithm. First, these functions are presented with the order of their recall in the algorithm, and at the end, the algorithm is coded. The required comments are presented. Note that the percent sign (%) represents a comment.

The first function is the objective function evaluation (*fobj*) which was presented in Chap. 2.

The second function is named the *Weight* function to determine the weights of the teams of the league (*W*). The input argument is the penalized objective function vector (*PFit*). The output argument is the weight vector (*W*).

```
% Define the weights of the teams of the league.
function [W]=Weight(PFit)
nT=size(PFit,2);

for i=1:nT
W(i)=(PFit(i)-min(PFit))/(max(PFit)-min(PFit))+1;
end
```

The third function is named *Rope_Pulling* function in which the total displacement of teams will be determined by a series of rope pulling competitions. The input arguments are the current positions of the teams or the league (*T*), the weight vector of the league (*W*), lower and upper bound vectors of the design variables (*Lb* and *Ub*), static and kinematic coefficients of friction (μ_s and μ_k), controlling parameters (α and β) and the time step parameter (ΔT) of the step size equation, and the current number of algorithm iterations (*nIT*). The output argument is matrix of the total displacement of teams (*stepsize*). The *Rope_Pulling* function is coded as follows:

```
% Determine the total displacement of teams by a series of rope pulling
competitions.

function
[stepsize]=Rope_Pulling(T,W,Lb,Ub,meus,meuk,deltat,alpha,beta,NITs)

nT=size(T,1);
nV=size(T,2);

stepsize=zeros(size(T));

for i=1:nT
    for j=1:nT
        if W(i)<W(j)
            PF=max(W(i)*meus,W(j)*meus); % Pulling force between teams i
and j.
            RF=PF-W(i)*meuk; % The resultant force affecting the team i due
to its interaction with the heavier team j.
            g=T(j,:)-T(i,:); % Gravitational acceleration.
            a=RF/(W(i)*meuk).*g; % Acceleration of team i towards team j.
            stepsize(i,:)=stepsize(i,:)+0.5*a*deltat^2+alpha^NITs*beta*(Lb-
Ub).*randn(1,nV);
        end
    end
end
```

The fourth function, named as *BTS* function by which the updated teams exited from the search space, regenerated using the best team strategy (Rule 5). Input arguments are the updated set of teams (*newt*), the current set of the teams of the league (*T*), the so far best-monitored team (*bestT*), lower and upper bound vectors of the design variables (*Lb* and *Ub*), and the current number of algorithm iterations (*nIT*). The output argument is the updated set of teams (*newt*) with corrected positions. The *BTS* function is coded as follows:

```
% Regenerate the teams exited from the search space using the Best Team
Strategy (BTS).

function [newT]=BTS(newT,T,bestT,Lb,Ub,NITs)

nT=size(newT,1);
nV=size(newT,2);

for i=1:nT
    for j=1:nV
        if newT(i,j)<Lb(j)||newT(i,j)>Ub(j)
            if rand<=0.5
            newT(i,j)=bestT(j)+randn/NITs*(bestT(j)-newT(i,j)); % Best Team
Strategy.
            if newT(i,j)<Lb(j)||newT(i,j)>Ub(j)
                newT(i,j)=T(i,j); % If the component is still outside the
search space, return it using the flyback strategy.
            end
            else
                if newT(i,j)<Lb(j)
                newT(i,j)=Lb(j);
                end
                if newT(i,j)>Ub(j)
                    newT(i,j)=Ub(j);
                end
            end
        end
    end
end
```

The fifth function is the *Replacement* function by which the league updated comparing the new teams with current teams of league. The input arguments are current league (*T*), its corresponding objective (*Fit*) and penalized objective (*PFit*) function vectors, updated set of teams (*newt*) and their corresponding objective (*newFit*) and penalized objective (*newPFit*) function vectors. The output arguments are updated league and its corresponding objective and penalized objective function vectors. *Replacement* function is coded as follows:

```
% Updating the league by comparing the updated teams with current teams of
league.

function [T,Fit,PFit]=Replacement(T,Fit,PFit,newT,newFit,newPFit)
nT=size(T,1);

helpT=[T;newT];helpFit=[Fit newFit];helpPFit=[PFit newPFit];

[~,order]=sort(helpPFit);

T=helpT(order(1:nT),:);
Fit=helpFit(order(1:nT));
PFit=helpPFit(order(1:nT));
```

TWO algorithm is coded in the following composed of three parts: initialization, algorithm cyclic body, and monitoring the results.

```
clc
clear

%% Initialization

% Define the properties of COP (tension/compression spring design problem).
nV=3; % Number of design variables.
Lb=[0.05 0.25 2]; % Lower bounds of design variables.
Ub=[2 1.3 15]; % Upper bounds of design variables.

% Define the parameters of TWO algorithm.
nT=10; % Number of Teams.
meus=1;meuk=1; % Static and Kinematic coefficients of friction.
deltat=1; % Time step in the movement equation with a constant
acceleration.
alpha=0.99; % Controlling parameter of the algorithm's randomness.
beta=0.1; % Scaling factor of team's movement.
maxNFEs=20000; % Maximum Number of Objective Function Evaluations.

%Generate random initial solutions.
for i=1:nT
    T(i,:)=Lb+(Ub-Lb).*rand(1,nV); % Teams matrix or the league or matrix
of the initial candidate solutions or the initial population.
end

% Evaluate initial population (League) calling the fobj function
constructed in the second chapter and form its corresponding vectors of
objective function (Fit) and penalized objective function (PFit). It should
be noted that the design vectors all are inside the search space.
for i=1:nT
    [X,fit,pfit]=fobj(T(i,:),Lb,Ub);
    T(i,:)=X;
    Fit(i)=fit;
    PFit(i)=pfit;
end

% Monitor the best candidate solution or team (bestT) and its corresponding
objective function (minFit) and penalized objective function (minPFit).
[minPFit,m]=min(PFit);
minFit=Fit(m);
bestT=T(m,:);

%% Algorithm Body

NFEs=0; % Current Number of Objective Function Evaluations used by the
algorithm until yet.
NITs=0; % Number of algorithm iterations

while NFEs<maxNFEs
    NITs=NITs+1; % Update the number of algorithm iterations.

    % Define the weights of the teams of the league.
    W=Weight(PFit);

    % Determine the total displacement of teams by a series of rope pulling
competitions.
    [stepsize]=Rope_Pulling(T,W,Lb,Ub,meus,meuk,deltat,alpha,beta,NITs);

    % Update the teams in the league.
    newT=T+stepsize;

    % Regenerate the teams exited from the search space using the Best Team
Strategy (BTS).
```

```
    [newT]=BTS(newT,T,bestT,Lb,Ub,NITs);

    % Evaluate the updated teams calling the fobj function. It should be
noted
    % that the existed dimensions are corrected before within the BTS
function. However, to do not change the form of fobj function it is checked
again here which is not needed.
    for i=1:nT
        [X,fit,pfit]=fobj(newT(i,:),Lb,Ub);
        newT(i,:)=X;
        newFit(i)=fit;
        newPFit(i)=pfit;
    end
    NFEs=NFEs+nT;

    % Update the league by the replacement strategy.
    [T,Fit,PFit]=Replacement(T,Fit,PFit,newT,newFit,newPFit);

    % Monitor the best candidate solution (bestT) and its corresponding
objective function (minFit) and penalized objective function (minPFit).
    minPFit=PFit(1);minFit=Fit(1);bestT=T(1,:);


    % Display desired information of the iteration.
    disp(['NITs= ' num2str(NITs) '; minFit = ' num2str(minFit) '; minPFit
= ' num2str(minPFit)]);

    % Save the required results for post processing and visualization of
the algorithm performance.
    output1(NITs,:)=[minFit,minPFit,NFEs];
    output2(NITs,:)=[min(PFit),max(PFit),mean(PFit)];
    output3(NITs,:)=[bestT,NFEs];
end

%% Monitoring the results
figure;
plot((1:1:NITs),output2(:,1),'g',(1:1:NITs),output2(:,2),'r--
',(1:1:NITs),output2(:,3),'b-.')
legend('min','max','mean');
xlabel('NITs');
ylabel('PFit');
```

11.4 Experimental Evaluation

TWO has seven parameters which are as follows: number of teams or the size of league (*nT*), static and kinematic coefficients of friction (μ_s and μ_k), controlling parameters (α and β) and the time step parameter (ΔT) used in the step size equation, and the maximum number of objective function evaluations (*maxNFEs*) as the stopping criteria of the algorithm.

Considering a large enough value for *maxNFEs* to ensure convergence of the algorithms is essential. This parameter almost is problem dependent and should be considered larger for complicated and larger problems. It is considered here equal to 20,000 for TWO like other chapters.

Static and kinematic coefficients of friction (μ_s and μ_k) and the time step parameter (ΔT) have no essential effect on the algorithm performance and are considered equal to 1 as recommended before.

The number of algorithm individuals in the population-based metaheuristics can be a very important parameter that influences the algorithm performance. Figure 11.3 shows the convergence history for a single run of TWO algorithm from a same initial population with a different number of teams (5–100). In these runs, α and β parameters are considered as 0.99 (the largest value of the recommended range) and 0.1 (the smallest value of the recommended range), respectively, to deep search with small step sizes. According to this figure and statistical results (not presented here) obtained from more independent runs, values of nT between 10 and 30 are efficient in making a balance between exploration and exploitation. It is worth to mention that TWO gets trapped in local optima when the nT is less than 10. On the other hand, larger values lead to a reduction of the convergence rate and accuracy.

Simulation results show that controlling parameters (α and β) used in the step size equation are the most important parameters of TWO. The recommended values and their role in the algorithm performance are completely observed. Especially the α parameter with the values outside the recommended range completely interrupts efficiency of the algorithm.

At the end it is worth to mention that although TWO defines seven parameters in its framework, however, as it is discussed above, TWO has only two easily tunable parameters (α and β) in addition to the number of algorithm agents and the maximum number of objective function evaluations as the common parameters of all metaheuristics. Therefore, TWO has a great potential in solving many COPs because of its simple structure, easy to be tuned for the problem at hand, and its special formulation in such a way that considers the qualities of both of the candidate solutions in updating the population.

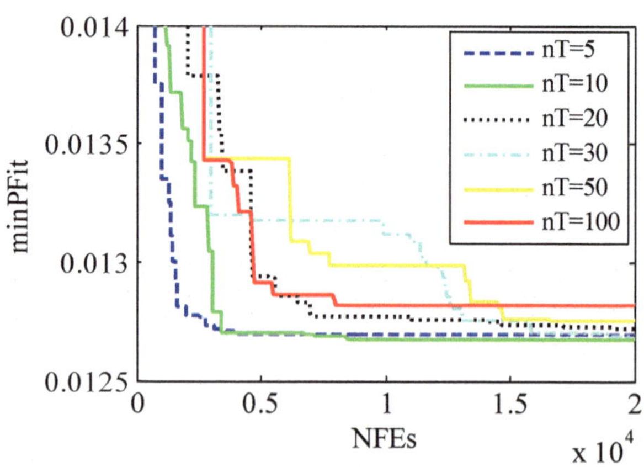

Fig. 11.3 Convergence histories of the TWO for different values of nT

References

1. Kaveh A, Zolghadr A (2016) A novel meta-heuristic algorithm: tug of war optimization. Int J Optim Civ Eng 6:469–492
2. Kaveh A, Shokohi F (2016) Optimum design of laterally-supported castellated beams using tug of war optimization algorithm. Struct Eng Mech 58:533–553
3. Kaveh A, Bijari S (2017) Bandwidth, profile and wavefront optimization using PSO, CBO, ECBO and TWO algorithms. Iran J Sci Trans Civ Eng 41:1–12
4. Kaveh A, Zolghadr A (2017) A guided modal strain energy-based approach for structural damage identification using tug-of-war optimization algorithm. J Comput Civ Eng 31(4):04017016

Chapter 12
Water Evaporation Optimization Algorithm

12.1 Introduction

Inspired by evaporation of a tiny amount of water molecules on the solid surface with different wettability which can be studied by molecular dynamic simulations, Kaveh and Bakhshpoori [1] developed a novel metaheuristic called Water Evaporation Optimization (WEO). Until today WEO is used successfully in solving some engineering COPs [2–6].

The evaporation of water is very important in biological and environmental science. Based on the molecular dynamic simulations, it is well-known that, as the surface changed from hydrophobicity to hydrophility, the evaporation speed does not show a monotonically decrease from intuition but increases first and then decreases after reaching a maximum value. When the surface wettability of the substrate is not high enough, the water molecules accumulate into the form of a sessile spherical cap. The predominant factor that affects the evaporation speed is the geometry shape of the water congregation. Meanwhile, when the surface wettability of the substrate is high enough, the water molecules spread to a monolayer, and the geometric factor no longer affects much, and the energy barrier provided by the substrate instead of the geometry shape affects the evaporation speed.

WEO considers water molecules as algorithm individuals. Solid surface or substrate with variable wettability is reflected as the search space. Decreasing the surface wettability (substrate changed from hydrophility to hydrophobicity) reforms the water aggregation from a monolayer to a sessile droplet. Such a behavior is consistent with how the layout of individuals changes to each other as the algorithm progresses. Decreasing wettability of the surface can represent the decrease of objective function for a minimizing optimization problem. Evaporation flux rate of the water molecules is considered as the most appropriate measure for updating the individuals which its pattern of change is in good agreement with the local and global search ability of the algorithm and can help WEO to have significantly well-converged behavior and simple algorithmic structure.

A. Kaveh, T. Bakhshpoori, *Metaheuristics: Outlines, MATLAB Codes and Examples*, https://doi.org/10.1007/978-3-030-04067-3_12

12.2 Formulation and Framework of the WEO

In the following firstly the results of molecular dynamic (MD) simulations for water evaporation from a solid surface as the essence of the WEO are introduced. Then the analogy between this physical phenomenon and a population-based metaheuristic is captured, and the framework of the WEO is presented. At the end, the steps of the WEO algorithm, its pseudo code, and the related flowchart are presented.

12.2.1 MD Simulations for Water Evaporation from a Solid Surface

Evaporation of water restricted on the surface of solid materials is the inspiration basis of WEO which is different from the water evaporation of bulk surface. This type of water evaporation is essential in the macroscopic world such as the water loss through the surface of the soil. Molecular dynamic (MD) simulations on the evaporation of water from a solid substrate with different surface wettability can be carried out by adhering nanoscale water aggregation in a neutral substrate which is chargeable [7]. By varying the value of charge (0 e $\leq q \leq$ 0.7 e), a substrate with tunable surface wettability can be obtained. Figure 12.1 depicts the MD simulation method: (a) side view of the initial system (the upward arrow denoted the

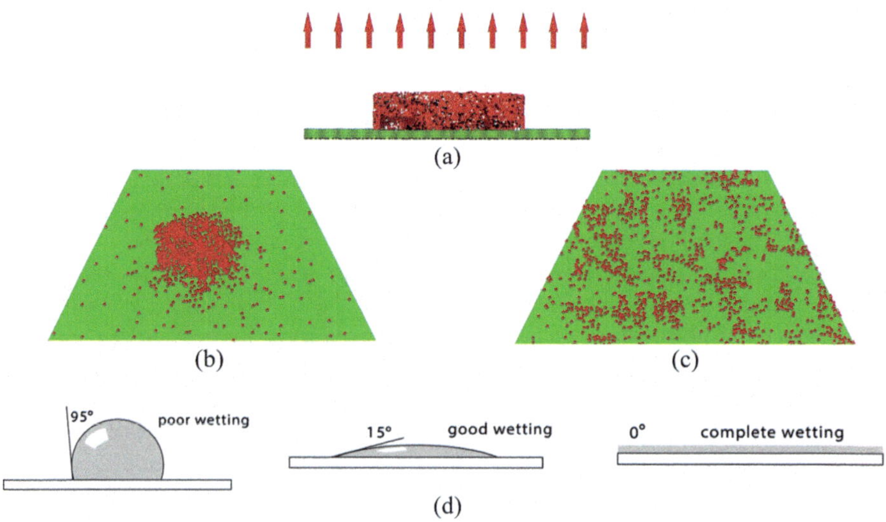

Fig. 12.1 (**a**) Side view of the initial system; (**b**) snapshot of water on the substrate with low wettability ($q = 0$ e); (**c**) snapshot of water on the substrate with high wettability ($q = 0.7$ e); (**d**) theoretical topology of water molecules with respect to substrate wettability used for MD simulations

accelerating region); (b) snapshot of water on the substrate with low wettability (the water molecules accumulate into the form of a sessile spherical cap with a contact angle θ to the surface); (c) snapshot of water on the substrate with high wettability (the adhered water forms a flat single-layer molecule sheet); and (d) theoretical topology of water molecules with respect to substrate wettability used for MD simulations.

The evaporation speed of the water layer can be described by the evaporation flux which is defined as the average number of the water molecules entering the accelerating region (the upward arrow denoted in Fig. 12.1a) from the substrate per nanosecond. Counter to intuition, as the surface changed from hydrophobicity ($q < 0.4$ e) to hydrophility ($q \geq 0.4$ e), the evaporation speed did not show a monotonically decrease but increased first and then decreased after it reached a maximum value.

To analyze this unusual variation of the evaporation flux, it can be assumed as a product of the aggregation probability of a water molecule in the interfacial liquid-gas surface and the escape probability of such surficial water molecule:

$$J(q) \propto P_{geo}(\theta(q))\, P_{ener}(E) \tag{12.1}$$

Here, $P_{geo}(\theta)$ is the probability for a water molecule on the liquid-gas surface, which is a geometry-related factor and is calculable by:

$$P_{geo}(\theta) = P_0\left(\frac{2}{3} + \frac{\cos^3\theta}{3} - \cos\theta\right)^{-2/3}(1 - \cos\theta) \tag{12.2}$$

where P_0 is a constant function of water molecule diameter and total volume of molecules.

$P_{ener}(E)$ is the escape probability of a surficial water molecule. $E = E_{WW} + E_{sub}(q)$ is the average interaction energy exerted on the surficial water molecules. E_{ww} is the energy provided by the neighboring water molecules. $E_{sub}(q)$ represents the interaction energy from the substrate, mainly provided by the electrical charge q assigned on the substrate.

Based on the MD simulation results, the relationship between the assigned charge q and the contact angle of the water droplet can be depicted as Fig. 12.2a. The contact angle of the water droplet θ decreases as q increases and reaches 0° when $q = 0.4$ e. When $q < 0.4$ e, most of the surficial water molecules are relatively far from the substrate. According to Fig. 12.2b, the energy E_{sub} provided by the substrate does not change much, and its variation is negligible if compared to the E_{sub} of $q \geq 0.4$ e. At the same time, the E_{ww} provided by the neighboring water molecules almost keeps constant in the simulation.

Hence, for $q < 0.4$ e, the escape probability of a surficial water molecule ($P_{ener}(E)$) is nearly a constant. Therefore, the evaporation flux (Eq. 12.1) will be updated as follows in which J_0 is a constant equal to 1.24 ns^{-1}.

$$J(\theta) = J_0\, P_{geo}(\theta), \quad q < 0.4\, e \tag{12.3}$$

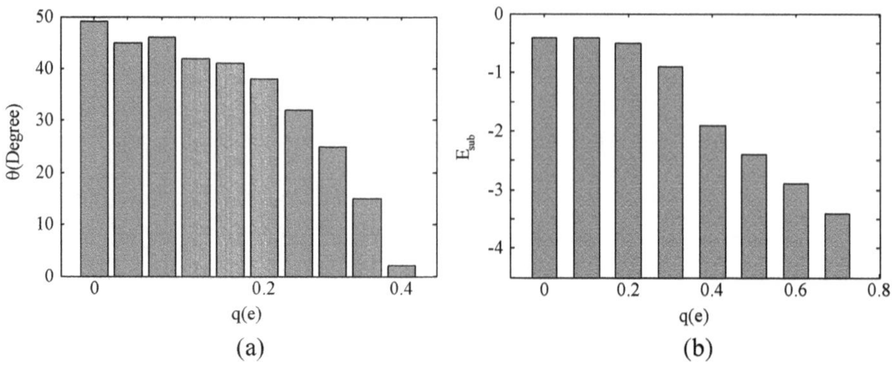

Fig. 12.2 (**a**) The contact angle θ of the water droplet with differently assigned charge q; (**b**) the interaction energy exerted on the outermost-layer water by the substrate (E_{sub}) with differently assigned charge q

For $q \geq 0.4$ e, the adhered water forms a flat single-layer molecule sheet with only a few water molecules overlapping upon it, and the shape of the tiny water aggregation does not change much with different q. According to the definition of $P_{geo}(\theta)$, all the water molecules are on the surface layer now; therefore $P_{geo}(\theta) = 1$. According to the thermal dynamics, for the system under the NVT ensemble (NVT ensemble indicates a canonical ensemble representing possible states of a mechanical system in thermal equilibrium), the probability for a free molecule to gain kinetic energy more than E_0 is proportional to $\exp\left(-\frac{E_0}{K_B T}\right)$. Based on the MD simulations, the evaporation flux decreases almost exponentially with respect to E_{sub}. Therefore, the evaporation flux (Eq. 12.1) will be updated as follows in which T is the room temperature and K_B is the Boltzmann constant.

$$J(q) = \exp\left(-\frac{E_{sub}}{K_B T}\right), \quad q \geq 0.4\, e \tag{12.4}$$

12.2.2 Analogy Between Water Evaporation and Population-Based Optimization

Considering MD simulation results overviewed in the previous subsection, from end to the beginning, a fine analogy can be found between this type of water evaporation phenomena and a population-based metaheuristic algorithm. This analogy is depicted in Fig. 12.3.

Water molecules and substrate with decreasing wettability are considered as algorithm individuals and search space, respectively. Decreasing the surface wettability reforms the water aggregation from a monolayer to a sessile droplet. Such a behavior is in coincidence with how the layout of the algorithm individuals changes to each other as the algorithm progresses. Decreasing q from 0.7 e to 0.0 e can

Water molecules	~	Algorithm individuals
Substrate with decreasing wettability	~	Search space with different objective function value
Water aggregation reforms from a monolayer to a sessile droplet with decreasing the surface wettability	~	Changing the layout of individuals to each other as the algorithm progresses
Decreasing q from 0.7 e to 0.0	~	Decreasing the objective function value
$q = 0.4$ e	~	the algorithm reaches the middle of the optimization process
Evaporation in two phases (Monolayer and droplet) with different evaporation flux	~	Global and local searchability of the algorithm

Fig. 12.3 The analogy between water evaporation from a solid surface and a population-based metaheuristic algorithm

represent the reduction of the objective function for a minimization optimization problem. Evaporation flux variation is considered as the most appropriate measure for updating the algorithm individuals which is in a good agreement with the local and global search ability.

Evaporation flux reaches its maximum around $q = 0.4$ e. This situation is considered in the WEO until the algorithm reaches the middle of the optimization process. In other words, WEO updates the individuals with the probability based on Eq. (12.4) (this probability is named monolayer evaporation probability or *MEP*) until it reaches to half the number of function evaluations. This first phase is considered as the global search ability of the algorithm. After this phase, individuals will be updated with the probability based on Eq. (12.3) (which is named as droplet evaporation probability or DEP). This phase is considered as the local searchability of the algorithm. These two phases and the updating mechanism of individuals are introduced at the following.

Monolayer Evaporation Phase

In the monolayer evaporation phase, Eq. (12.4) is estimated with a simple exponential function of the substrate interaction energy: exp (E_{sub}). In this phase ($q > 0.4$ e), as q increases, the substrate will have more energy, and as a result, less evaporation will be made. Based on Fig. 12.2b, WEO considers -0.5 and -3.5 as the maximum (E_{max}) and minimum (E_{min}) values of E_{sub}, respectively, for each iteration of the algorithm until half the number of algorithm iterations. Figure 12.4 shows the monolayer evaporation probability for various values of substrate energy between -3.5 and -0.5.

In each iteration, the penalized objective function of individuals *PFit(i)* is scaled to the interval $[-3.5, -0.5]$ and represents the corresponding $E_{sub}(i)$ inserted to each individual (substrate energy vector), via the following scaling function:

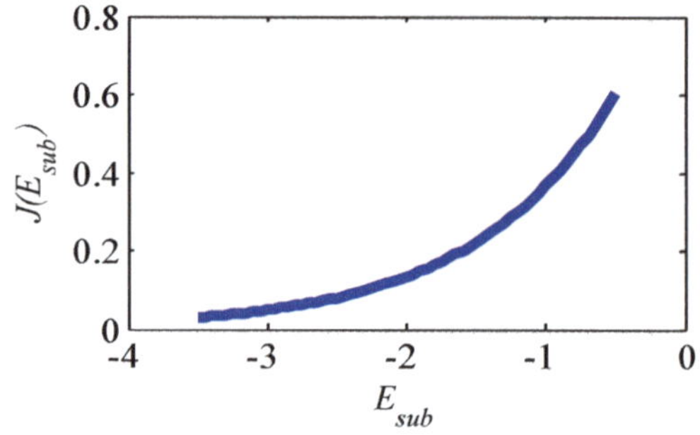

Fig. 12.4 Monolayer evaporation flux with different substrate energy for the WEO

$$E_{sub}(i) = \frac{(E_{max} - E_{min}) \times (PFit(i) - \mathrm{Min}(Fit))}{(\mathrm{Max}(Fit) - \mathrm{Min}(Fit))} + E_{min} \tag{12.5}$$

where *Min* and *Max* are the minimum and maximum functions, respectively. After generating the substrate energy vector, the Monolayer Evaporation Probability matrix (*MEP*) is constructed by the following equation:

$$MEP_{ij} = \begin{cases} 1 & \text{if } rand_{ij} < \exp(E_{sub}(i)) \\ 0 & \text{if } rand_{ij} \geq \exp(E_{sub}(i)) \end{cases} \tag{12.6}$$

where MEP_{ij} is the updating probability for the *j*th variable of the *i*th individual or water molecule. In this way, an individual with better objective function (considering the minimization problem) is more likely to remain unchanged in the search space. In detail we can say that in each iteration the best and worst candidate solutions will be updated by the probability equal to exp(−3.5) = 0.03 and exp (−0.5) = 0.6, respectively. WEO considers these values as a minimum (MEP_{min}) and maximum (MEP_{max}) values of monolayer evaporation probability. Simulation results show that considering $MEP_{min} = 0.03$ and $MEP_{max} = 0.6$ based on the simulation results (Fig. 12.2b) is logical. However, these values can be considered as the parameters of the algorithm.

Droplet Evaporation Phase

In the droplet evaporation phase, using Eqs. (12.2) and (12.3) the evaporation flux is as the following:

$$J(\theta) = J_0 P_0 \left(\frac{2}{3} + \frac{\cos^3\theta}{3} - \cos\theta\right)^{-2/3} (1 - \cos\theta) \tag{12.7}$$

where J_0 and P_0 are constant values. In this phase ($q < 0.4$ e), since *q* is smaller, the contact angle is greater and as a result, less evaporation will be accured. According to Fig. 12.2(a) the maximum and minimum values of contact angle are 50° and 0°, respectively. Based on the MD simulations results, the variation of the evaporation flux perfectly fitted to the experimental results in the range $20° < \theta < 50°$. It can be

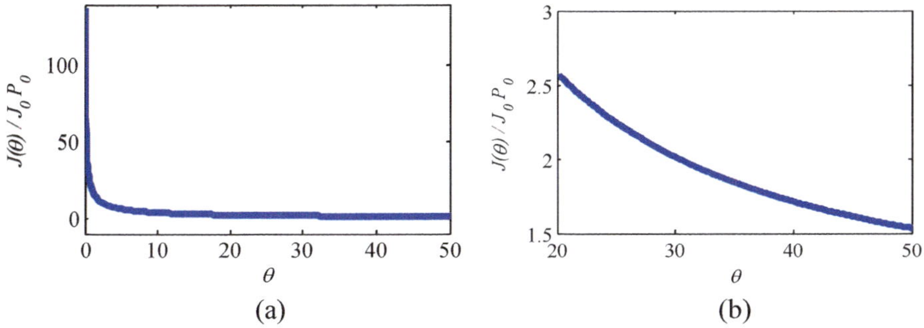

Fig. 12.5 Droplet evaporation flux based on the MD simulations

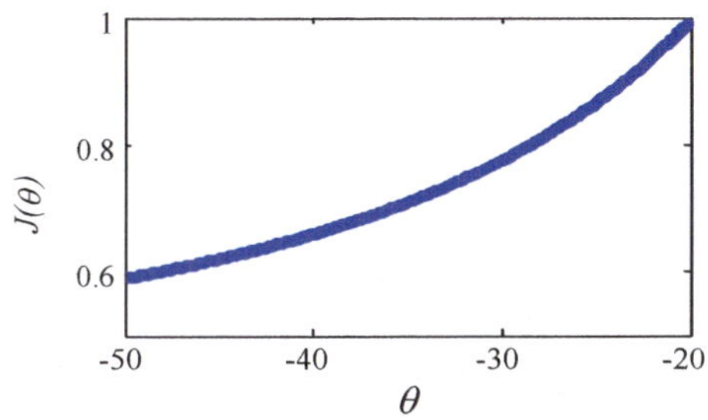

Fig. 12.6 Droplet evaporation flux with different contact angles considered for the WEO

interpreted that for $\theta < 20°$ the water droplet is no longer observed like a perfect sessile spherical cap. Figure 12.5a and b illustrate this evaporation flux functions neglecting the constant values J_O and P_O for various contact angles between $0° < \theta < 50°$ and $20° < \theta < 50°$, respectively.

Considering the contact angle between $20° < \theta < 50°$ is quite suitable for WEO. Based on Fig. 12.5b, the maximum value for droplet evaporation probability is 2.6. Considering $J_O \times P_O$ equal to $\frac{1}{2.6}$ for limiting the upper bound of droplet evaporation probability to 1, and considering -20 and -50 as the maximum (θ_{max}) and minimum (θ_{min}) values of the contact angle, the droplet evaporation probability for various contact angles between $-50° < \theta < -20°$ is shown in Fig. 12.6. For all iterations in the second half of the algorithm, the penalized objective function of individuals *PFit(i)* is scaled to the interval $[-50°, -20°]$ via the following scaling function which represents the corresponding contact angle $\theta(i)$ (contact angle vector):

$$\theta(i) = \frac{(\theta_{max} - \theta_{min}) \times (PFit(i) - \mathrm{Min}(Fit))}{(\mathrm{Max}(Fit) - \mathrm{Min}(Fit))} + \theta_{min} \tag{12.8}$$

where the *Min* and *Max* are the minimum and maximum functions. Such an assumption is consistent with MD simulations as depicted in Fig. 12.2a and results in a good performance of WEO. Negative values have no effect on computations (cosine is an even function). In this way, the best and worst individuals have the smaller and bigger updating probability like the evaporation speed of a droplet on a substrate with less ($q = 0.0$ e) and more ($q = 0.4$ e) wettability, respectively. In other words, we can have the droplet evaporation probability matrix with minimum (DEP_{min}) and maximum (DEP_{max}) values of droplet evaporation probability equal to 0.6 and 1, respectively, as shown in Fig. 12.6. Performance evaluation results show that these values are suitable for WEO. However, these parameters can be considered as the next two parameters of the algorithm.

After generating contact angle vector $\theta(i)$, the droplet evaporation probability (*DEP*) matrix is constructed by the following equation:

$$DEP_{ij} = \begin{cases} 1 & \text{if } rand_{ij} < J(\theta_i) \\ 0 & \text{if } rand_{ij} \geq J(\theta_i) \end{cases} \tag{12.9}$$

where DEP_{ij} is the updating probability for the *j*th variable of the *i*th individual or water molecule.

Updating Water Molecules

In the MD simulations, the number of evaporated water molecules in the entire simulation process is considered negligible compared to the total number of the water molecules resulting in a constant total number of molecules. In the WEO also the number of algorithm individuals or number of the water molecules (*nWM*) is considered constant, in all algorithm iterations. *nWM* is the algorithm parameter like other population-based algorithms. Considering a maximum value for algorithm iterations (*maxNITs*) is essential for WEO to determine the evaporation phase of the algorithm and also to use as the stopping criterion. Such stopping criterion is utilized in many of optimization algorithms. When a water molecule is evaporated, it should be renewed. Updating or evaporation of the current water molecules is made with the aim of improving objective function. The best strategy for regenerating the evaporated water molecules is using the current set of water molecules (*WM*). In this way a random permutation-based step size can be considered for possible modification of individuals as:

$$stepsize = rand.(WM[permute1(i)(j)] - WM[permute2(i)(j)]) \tag{12.10}$$

where *rand* is a random number in [0, 1] range and *permute*1 and *permute*2 are different rows permutation functions. *i* is the number of water molecule, and *j* is the number of dimensions of the problem at hand. The next set of molecules (*newWM*) is generated by adding this random permutation-based step size multiplied by the corresponding updating probability (monolayer evaporation and droplet evaporation probability) and can be stated mathematically as:

$$newWM = WM + stepsize \times \begin{cases} MEP & NITs \leq maxNITs/2 \\ DEP & NITs > maxNITs/2 \end{cases} \tag{12.11}$$

Each water molecule is compared and replaced by the corresponding renewed molecule based on the objective function. It should be noted that random permutation-based step size helps WEO in two aspects. In the first phase, water molecules are farther from each other than the second phase. In this way the generated permutation-based step size will guarantee global and local search capability in each phase. The random part will guarantee the algorithm to be sufficiently dynamic. It should also be noted that these two aspects are guaranteed with more emphasis by considering two specific evaporation probability mechanisms for each phase: in both evaporation phases, as it is clear from Figs. 12.4 and 12.6, best water molecules are renewed locally (with less evaporation probability), while bad-quality molecules are renewed globally (with more evaporation probability).

12.2.3 The WEO Algorithm

At the following WEO algorithm is organized in five steps, and then its pseudocode and flowchart are presented.

Step 1: Initialization
Algorithm parameters are set in the first step. These parameters are the number of water molecules (*nWM*) and maximum number of algorithm iterations (*maxNITs*). It should be noted that the minimum (MEP_{min}) and maximum (MEP_{max}) values of monolayer evaporation probability and the minimum (DEP_{min}) and maximum (DEP_{max}) values of droplet evaporation probability can also be considered as the algorithm parameters. However, the evaporation probability parameters are determined efficiently for WEO based on the MD simulations results ($MEP_{min} = 0.03$ and $MEP_{max} = 0.6$; $DEP_{min} = 0.6$ and $DEP_{max} = 1$). WEO starts from *nWM* number of candidate solutions or water molecules randomly generated within the search space. These solutions construct the matrix of water molecules (*WM*). After evaluating the molecules, the corresponding objective function (*Fit*) and the penalized objective function (*PFit*) vectors are produced.

Step 2: Generating Water Evaporation Matrix
Every water molecule follows the evaporation probability rules specified for each phase of the algorithm in the previous subsection.

Step 3: Generating Random Permutation-Based Step Size Matrix
A random permutation-based step size matrix is generated according to Eq. (12.10).

Step 4: Generating Evaporated Water Molecules and Updating the Matrix of Water Molecules
The evaporated set of water molecules *newWM* is generated by adding the product of step size matrix and evaporation probability matrix to the current set of molecules *WM* according to Eq. (12.11).

These new water molecules are evaluated based on the objective function. For the molecule i ($i = 1, 2, \ldots, nWM$), if the newly generated molecule i ($i = 1, 2, \ldots,$

nWM) is better than the old one, it will replace it. The best water molecule (*bestWM*) is returned.

Step 5: Terminating Condition
If the number of iteration of the algorithm (*NITs*) becomes larger than the maximum number of iterations (*maxNITs*), the algorithm terminates. Otherwise go to Step 2.

The pseudo code of WEO is given as follows, and the flowchart is illustrated in Fig. 12.7.

The pseudo code of the WEO algorithm for solving COPs:

Define the algorithm parameters: *nWM* and *maxNFEs*.
Generate random initial water molecules (*WM*).
Evaluate the initial molecules and form its corresponding vectors of the objective function (*Fit*) and penalized objective function (*PFit*).
While $NFEs \leq maxNFEs$

Update *NITs*.
if $NITs \leq maxNITs/2$

Generate new water molecules based on the monolayer evaporation strategy.
Evaluate the newly generated water molecules, and replace the current molecules with the evaporated ones if the newest ones are better.
Update *NFEs*.

else

Generate new water molecules based on the droplet evaporation strategy.
Evaluate the newly generated water molecules, and replace the current molecules with the evaporated ones if the newest ones are better.
Update *NFEs*.

end if
Determine and monitor the best water molecule (*bestWM*).

end While

12.3 Matlab Code for the Water Evaporation Optimization (WEO) Algorithm

According to the previous section, four functions are considered for coding the algorithm. First, these functions are presented with the order of their recall in the algorithm, and at the end, the algorithm is coded. The required comments are presented. Note that the percent sign (%) represents a comment.

The first function is the objective function evaluation (*fobj*) presented in Chap. 2.

The second function is named as the *Monolayer_Evaporation* function by which molecules are updated based on the monolayer evaporation strategy. The input arguments are the current position of the water molecules matrix (*WM*) and its corresponding penalized objective function vector (*PFit*). The output argument is newly generated molecules or the evaporated set of water molecules (*newWM*).

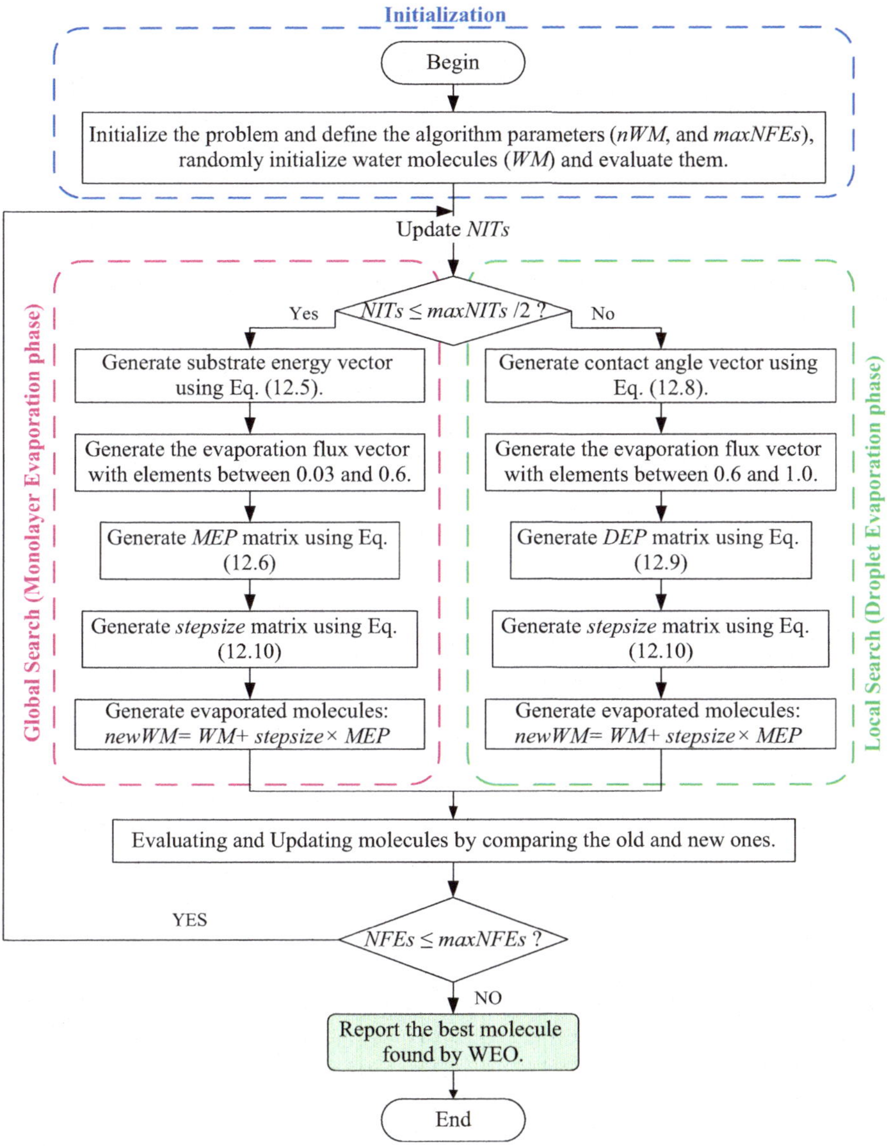

Fig. 12.7 Flowchart of the WEO algorithm

```
% Get new water molecules by monolayer evaporation strategy.

function newWM=Monolayer_Evaporation(WM,PFit)
Emax=-.5;Emin=-3.5; % Maximum and Minimum values of the substrate energy.
for i=1:size(PFit,2)
    E(i)=(Emax-Emin)*(PFit(i)-min(PFit))/(max(PFit)-min(PFit))+Emin;         %
Substrate energy vector.
    Jesub(i)=exp(E(i));  %  The  evaporation  flux  vector  (with  elements
between 0.03 and 0.6).
    MEP(i,:)=rand(1,size(WM,2))<Jesub(i);     %     Monolayer     Evaporation
Probability matrix.
end
stepsize=rand*(WM(randperm(size(WM,1)),:)-WM(randperm(size(WM,1)),:));
newWM=WM+stepsize.*MEP;
```

The third function is the *Droplet_Evaporation* function in which molecules are updated based on the droplet evaporation strategy. The input and output arguments are same as the previous function.

```
%Replace some nests by constructing new solutions/nests

function newWM=Droplet_Evaporation(WM,PFit)

Tetamax=-20;Tetamin=-50; % The maximum and minimum values of the contact
angle.
for i=1:size(PFit,2)
    Teta(i)=(Tetamax-Tetamin)*(PFit(i)-min(PFit))/(max(PFit)-
min(PFit))+Tetamin; % Contact angle vector.
    Jteta(i)=(1/2.6)*(2/3+(cosd(Teta(i)))^3/3-cosd(Teta(i)))^(-2/3)*(1-
cosd(Teta(i))); % The evaporation flux vector (with elements between  0.6
and 1.0).
    DEP(i,:)=rand(1,size(WM,2))<Jteta(i); % Droplet Evaporation Probability
matrix.
end

stepsize=rand*(WM(randperm(size(WM,1)),:)-WM(randperm(size(WM,1)),:));
newWM=WM+stepsize.*DEP;
```

The fourth function is the *Replacement* function by which the newly generated water molecules evaluated and compared with the current set of molecules, so that the current bad molecules are replaced with the newly generated better ones. It should be noted that the *fobj* function is recalled in this function in the nested form. Input arguments are the current set of (*WM*) and newly generated water molecules (*newWM*), corresponding objective function (*Fit*) and the penalized objective function (*PFit*) vectors of the current set of the water molecules, and the lower (*Lb*) and upper (*Ub*) bounds of design variables. The *Replacement* function is coded as follows:

```
% Evaluating and Updating molecules by comparing the old and new ones.

function [WM,Fit,PFit]=Replacemnet(WM,newWM,Fit,PFit,Lb,Ub)

for i=1:size(WM,1),
     [X,fit,pfit]=fobj(newWM(i,:),Lb,Ub);
     if pfit<=PFit(i)
         WM(i,:)=X;
         Fit(i)=fit;
         PFit(i)=pfit;
     end
end
```

The WEO algorithm is coded in the following, and it is composed of three parts: initialization, algorithm cyclic body, and monitoring the results.

```
clc
clear
% rng('default');
%% Initialization

% Define the properties of COP (tension/compression spring design problem).
nV=3; % Number of design variables.
Lb=[0.05 0.25 2]; % Lower bounds of design variables.
Ub=[2 1.3 15]; % Upper bounds of design variables.

% Define the parameters of the WEO algorithm.
nWM=20; % Number of Water Molecules.
maxNFEs=20000; % Maximum Number of Objective Function Evaluations.

% Randomly generate nWM number of water molecules as the initial population
of the algorithm.
for i=1:nWM
    WM(i,:)=Lb+(Ub-Lb).*rand(1,nV); % Thermal objects matrix or matrix of
the initial candidate solutions or the initial population..
end

% Evaluate initial population (WM) calling the fobj function constructed in
% the second chapter and form its corresponding vectors of objective
function (Fit) and penalized objective function (PFit). It should be noted
that the design vectors all are inside the search space.
for i=1:nWM
    [X,fit,pfit]=fobj(WM(i,:),Lb,Ub);
    WM(i,:)=X;
    Fit(i)=fit;
    PFit(i)=pfit;
end

%% Algorithm Body

NFEs=0; % Current Number of Objective Function Evaluations used by the
algorithm until yet.
NITs=0; % Number of algorithm iterations.
maxNITs=maxNFEs/nWM; % Maximum Number of algorithm iterations.

while NFEs<maxNFEs
    NITs=NITs+1; % Update the number of algorithm iterations.

    if NITs<=maxNITs/2
        % Generate new water molecules based on the monolayer evaporation
        % strategy.
        newWM=Monolayer_Evaporation(WM,PFit);

        % Replace the current molecules with the evaporated ones if the
newest ones are better. Notice that the fobj function is called in the
following replacement function in nested form. Hence the newly generated
molecules will be corrected and evaluated.
        [WM,Fit,PFit]=Replacemnet(WM,newWM,Fit,PFit,Lb,Ub);

        % Update the number of objective function evaluations used by the
algorithm until yet.
        NFEs=NFEs+nWM;
    else
        % Generate new water molecules based on the droplet evaporation
strategy.
        newWM=Droplet_Evaporation(WM,PFit);
```

```
        % Replace the current molecules with the evaporated ones if the
newest ones are better. Notice that the fobj function is called in the
following replacement function in nested form. Hence the newly generated
molecules will be corrected and evaluated.
        [WM,Fit,PFit]=Replacemnet(WM,newWM,Fit,PFit,Lb,Ub);

        % Update the number of objective function evaluations used by the
algorithm untill yet.
        NFEs=NFEs+nWM;
    end

    % Monitor the best candidate solution (bestWM) and its corresponding
penalized objective funtion (minPFit) and objective function (minFit).
    [minPFit,m]=min(PFit);
    minFit=Fit(m);
    bestWM=WM(m,:);

    % Display desired information of the iteration.
    disp(['NITs= '  num2str(NITs) '; minFit = '  num2str(minFit) '; minPFit
= '  num2str(minPFit)]);

    %Save the required results for post processing and visualization of
algorithm performance.
    output1(NITs,:)=[minFit,minPFit,NFEs];
    output2(NITs,:)=[min(PFit),max(PFit),mean(PFit)];
    output3(NITs,:)=[bestWM,NFEs];
end

%% Monitoring the results
figure;
plot((1:1:NITs),output2(:,1),'g',(1:1:NITs),output2(:,2),'r--
',(1:1:NITs),output2(:,3),'b-.')
legend('min','max','mean');
xlabel('NITs');
ylabel('PFit');
```

12.4 Experimental Evaluation

A very significant point which attracts attention is that WEO has just two parameters which are common for all population-based metaheuristics: the number of population and the maximum number of objective function evaluations as the stopping criteria. Four other parameters of controlling the evaporation probability matrices are determined based on the MD simulations results.

Kaveh and Bakhshpoori [1, 2] conducted a deep parametric and sensitivity analysis of the WEO algorithm to study its search behavior. The readers are referred to these manuscripts for more detail. Simulation results show that considering at least seven numbers of water molecules is essential for the WEO and the larger number will cause more computational cost.

References

1. Kaveh A, Bakhshpoori T (2016a) Water evaporation optimization: a novel physically inspired optimization algorithm. Comput Struct 167:69–85
2. Kaveh A, Bakhshpoori T (2016b) A new metaheuristic for continuous structural optimization: water evaporation optimization. Struct Multidiscip Optim 54:23–43
3. Kaveh A, Bakhshpoori T (2016c) An accelerated water evaporation optimization formulation for discrete optimization of skeletal structures. Comput Struct 177:218–228
4. Saha A, Das P, Chakraborty AK (2017) Water evaporation algorithm: a new metaheuristic algorithm towards the solution of optimal power flow. Eng Sci Technol Int J 20(6):1540–1552
5. Venkadesh R, Anandhakumar R (2017) Economic dispatch with multiple fuel options using water evaporation optimization. Int J Comput Appl 165:29–35
6. Venkadesh R, Radhakrishnan A (2017) Water evaporation algorithm to solve combined economic and emission dispatch problems. Glob J Pure Appl Math 13:1049–1067
7. Wang S, Tu Y, Wan R, Fang H (2012) Evaporation of tiny water aggregation on solid surfaces with different wetting properties. J Phys Chem B 116(47):13863–13867

Chapter 13
Vibrating Particles System Algorithm

13.1 Introduction

Vibrating Particles System (VPS) algorithm is a new metaheuristic search algorithm developed by Kaveh and Ilchi Ghazaan [1]. VPS is motivated based on the free vibration of single degree of freedom systems with viscous damping. VPS has been used to solve some problems in the field of structural optimization, and the obtained results show its viability in convergence and accuracy [2–5].

Like other population-based metaheuristics, VPS starts from a random set of initial solutions and considers them as the free vibrated single degree of freedom systems with viscous damping. Considering under-damped conditions, each free vibrated system or vibrating particle will oscillate and return to its equilibrium position with a specific formulation, which is easily verifiable using differential equations. By utilizing a combination of randomness and exploitation of the obtained results, VPS improves the quality of the particles iteratively by oscillating them toward the equilibrium position, as the optimization process proceeds. Consider the equilibrium position of each particle, consisting of three parts, the best position achieved so far across the entire population (HP), a good particle (GP), and a bad particle (BP). In this way the essence of VPS stands on three essential concepts, self-adaptation (particle moves toward HB), cooperation (the GP and BP, which are selected from particles themselves, can influence the new position of particles), and competition (the influence of GP will be more than that of BP). It should be noted that VPS uses a memory based on the harmony search strategy for correcting the position of particles exited from the search space.

A. Kaveh, T. Bakhshpoori, *Metaheuristics: Outlines, MATLAB Codes and Examples*, https://doi.org/10.1007/978-3-030-04067-3_13

13.2 Formulation and Framework of the VPS

In the following, the concepts and framework of VPS algorithm, its pseudo code, and the related flowchart are presented. The formulation of the free vibration of an under-damped single degree of freedom (SDOF) system [1] is not provided here extensively because of its simplicity and not using it directly in the formulation of VPS. However, it is firstly outlined at the following.

The equation of motion for a damped free vibration of a SDOF system in terms of x (degree of freedom), m (mass), c (damping coefficient), and k (spring constant) is as follows:

$$m\ddot{x} + c\dot{x} + kx = 0 \tag{13.1}$$

The solution of this equation for the under-damped conditions is as follows:

$$x(t) = \rho\, e^{-\xi\omega_n t} \sin\left(\omega_D t + \phi\right) \tag{13.2}$$

where ρ and ϕ are constants generally determined from the initial conditions of the vibration and ω_n and ξ are damped natural frequency and damping ratio, respectively. This solution is depicted in Fig. 13.1. As it is clear, the free vibration of an under-damped SDOF oscillates and returns to its rest or equilibrium position. And this is the point by which the VPS is inspired.

Like other population-based metaheuristics, VPS starts from a set of candidate solutions randomly generated within the search space. The number of vibrating particles is considered as nVP. These particles form the matrix of Vibrating Particles (VP). After evaluating the objects, the corresponding objective function (Fit) and the penalized objective function ($PFit$) are produced.

VPS updates the particles in a way that considers for each particle, three equilibrium positions with different weights (ω_1, ω_2, and ω_3) that the particle tends to approach: (1) the best position achieved so far across the entire population (HP), (2) a good particle (GP), and (3) a bad particle (BP). In order to select GP and BP for each particle, the current population is sorted according to their penalized objective

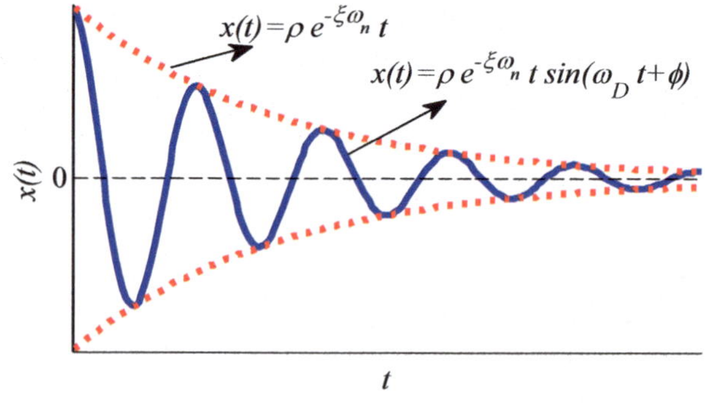

Fig. 13.1 The free vibrating motion of an under-damped SDOF

function values in an increasing order, and then GP and BP are chosen randomly from the first and second halves except itself, respectively.

Damping level plays an important role in the vibration. Much more damping level higher rater at which the amplitude of a free damped vibration decrease. In order to model this phenomenon in the VPS, a descending function (*D*) proportional to the number of iterations is proposed as follows:

$$D = (NITs/maxNITs)^{-\alpha} \tag{13.3}$$

where *NITs* is the current iteration number of the algorithm, *maxNITs* is the maximum number of algorithm iterations considered as the stopping criteria, and α is a constant. The value of 0.05 is recommended for it.

According to the mentioned concepts, the particles are updated by the following formula which will be read as free vibration formula hereafter:

$$\begin{aligned} newVP_i &= \omega_1(D.A.rand + HP) + \omega_2(D.A.rand + GP_i) \\ &\quad + \omega_3(D.A.rand + BP_i), \\ A &= \omega_1(HP - VP_i) + \omega_2(GP_i - VP_i) + \omega_3(BP_i - VP_i), \\ &\omega_1 + \omega_2 + \omega_3 = 1 \end{aligned} \tag{13.4}$$

in which VP_i and $newVP_i$ are the current and updated positions of the *i*th particle, respectively; ω_1, ω_2, and ω_3 are three weights to measure the relative importance of the best-so-far particle found by the algorithm (*HP*), the good particle (*GP*), and bad particle (*BP*) of the *i*th particle, respectively; and *rand*s are random numbers uniformly generated between zero and one. The effects of *A* and *D* functions are similar to those of ρ and $e^{-\xi\omega_n t}$ in Eq. (13.2), respectively. Also, the value of sin ($\omega_D t + \phi$) is considered as unity.

A parameter like *p* within (0, 1) is defined, and it is specified whether the effect of *BP* must be considered in updating position or not. For each particle, *p* is compared with rand (a random number uniformly distributed in the range of [0,1]); if $p <$ rand, then $\omega_3 = 0$ and $\omega_2 = 1 - \omega_1$.

Three essential concepts, consisting of self-adaptation, cooperation, and competition, are considered in VPS. A particle moves toward HP, so the self-adaptation is provided. Any particle has the chance to have an influence on the new position of the other one, so the cooperation between the particles is supplied. Due to the *p* parameter, the influence of GP (good particle) is more than that of BP (bad particle); therefore, the competition is provided.

As it was mentioned in the introduction section, VPS uses harmony search-based handling approach to deal with a particle violating the limits of the variables. This approach has previously been addressed in detail within Chap. 7 (Charged System Search Algorithm). However, to maintain the consistency of the current chapter, it is presented here again. A vibrating particles memory (*VP-M*) is utilized to save the *nVP* number of the best vibrating particles and their related objective function (*Fit-M*) and penalized objective function (*PFit-M*) values. To fulfill this aim, vibrating particles memory is utilized to save the same number with the number of the particles (*nVP*). Considering memory and benefitting it in the form of different strategies can improve the metaheuristics performance, without increasing the computational cost. It should be noted again that VPS used it just for regenerating the particles exited from the search space. According to this mechanism, any component of the solution vector violating the variable boundaries can be regenerated from the *VP-M* as:

$$VP(i,j) = \begin{cases} \text{w.p.}\,vpmcr & \Rightarrow \text{ select a new value for a variable from } VP\text{-}M, \\ & \Rightarrow \text{ w.p.}(1-par) \text{ do nothing,} \\ & \Rightarrow \text{ w.p.}\,par \text{ choose a neighboring value,} \\ \text{w.p.}(1-vpmcr) & \Rightarrow \text{ select a new value randomly,} \end{cases} \tag{13.5}$$

where "w.p." is the abbreviation for "with the probability," *VP(i, j)* is the *j*th component of the *i*th vibrating particle, *vpmcr* is the vibrating particle memory considering rate varying between 0 and 1 and sets the probability of choosing a value in the new vector from the historic values stored in *VP-M*, and $(1 - vpmcr)$ sets the probability of choosing a random value from the possible range of values. The pitch-adjusting process is performed only after a value is chosen from *VP-M*. The value $(1 - par)$ sets the rate of doing nothing, and *par* sets the rate of choosing a value from neighboring the best vibrating particle or the particles saved in memory. For choosing a value from neighboring the best vibrating particle or the particles saved in memory, for continuous search space, a randomly generated step size can be used ($\pm bw \times \text{rand}$).

The pseudo code of VPS is given as follows, and the flowchart is illustrated in Fig. 13.2.

The pseudo code of VPS algorithm for solving COPs:

Define the algorithm parameters: *nVP*, α, ω_1, ω_2, *p*, *vpmcr*, *par*, *bw*, and *maxNFEs*.
Generate random initial vibrating particles (*VP*).
Evaluate the initial vibrating particles and form its corresponding vectors of the objective function (*Fit*) and penalized objective function (*PFit*).
Form the vibrating particles memory matrix (*VP_M*) and its corresponding vectors of objective function memory (*Fit_M*) and penalized objective function memory (*PFit_M*).
Determine the best particle obtained sofar by the algorithm (*HP*).
While *NFEs* < *maxNFEs*:

Update the number of algorithm iterations (*NITs*).
Determine the *D* parameter using Eq. (13.3).
Determine the good particle (GP) and bad particle (BP) matrixes.
Update particles based on the Eq. (13.4).
Regenerate the vibrating particles exited from the search space using the harmony search-based handling approach.
Evaluate the new vibrating particles.

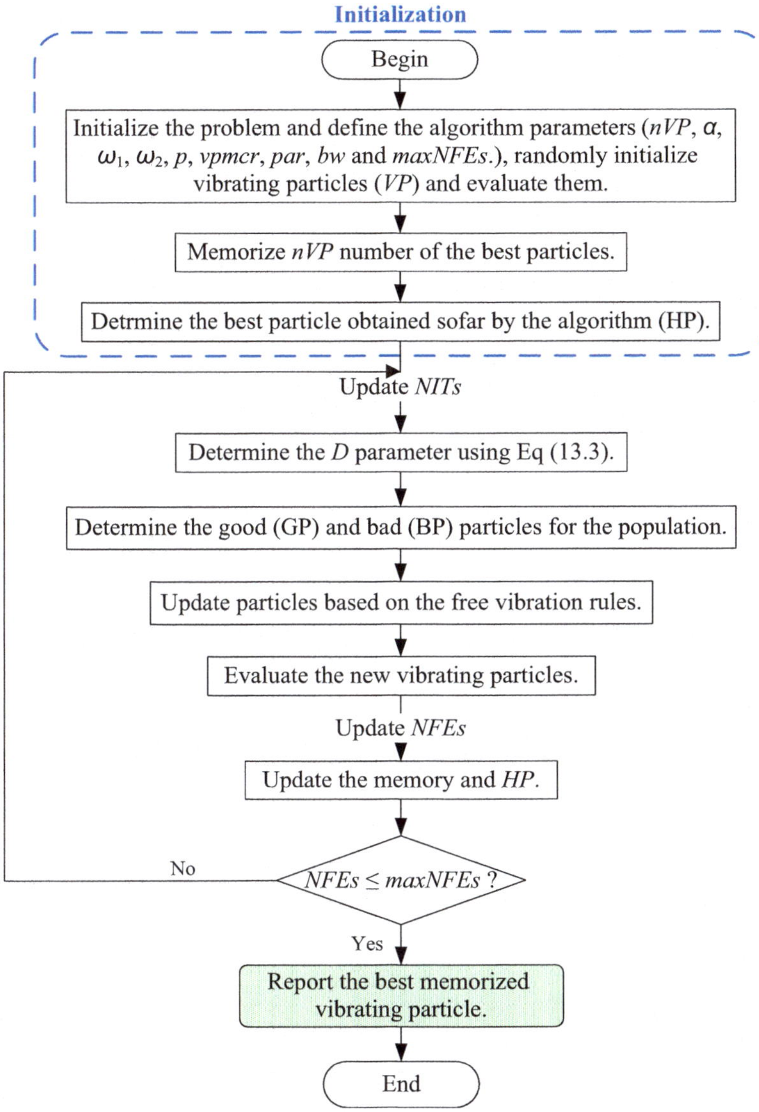

Fig. 13.2 Flowchart of the VPS algorithm

Update *NFEs*.

Update the vibrating particles memory matrix (*VP_M*) and its corresponding vectors of objective function memory (*Fit_M*) and penalized objective function memory (*PFit_M*).

Determine and monitor the best memorized candidate solution (*HP*).

end While

13.3 MATLAB Code for the Vibrating Particle System (VPS) Algorithm

According to the previous section, five functions are considered for coding the algorithm. First, these functions are presented with the order of their recall in the algorithm, and at the end, the algorithm is coded. The required comments are presented. Note that the percent sign (%) represents a comment.

The first function is the objective function evaluation (*fobj*) which was presented in Chap. 2.

The second function named as the *Particles* function to determine the good particles (GP) and bad particles (BP) matrixes for updating the population. The input arguments are the current position of the vibrating particles matrix (*VP*) and their corresponding penalized objective function (*PFit*). The output arguments are *GP* and *BP* matrixes.

```
% Determine the good particles (GP) and bad particles (BP) matrixes.
function [GP,BP]=Particles(VP,PFit)

nVP=size(VP,1);

% For this purpose firstly sort the current population based on the merits.
Select GP and BP randomly for each candidate solution from the first and
second halves except itself, respectively.

[~,order]=sort(PFit);

for i=1:nVP
    iGP=i; % Index of the good particle
    while iGP==i
        iGP=order(ceil(rand*0.5*nVP));
        GP(i,:)=VP(iGP,:);
    end
    iBP=i; % Index of the bad particle
    while iBP==i
        iBP=order(nVP-ceil(rand*0.5*nVP)+1);
        BP(i,:)=VP(iGP,:);
    end
end
```

The third function is the *Free_Viberation* function in which the vibrating particles updated to a new position of equilibrium. The input arguments are *VP*, *HP*, *GP*, *BP*, ω_1, ω_2, ω_3, and *p*. The output argument is the updated vibrating particles matrix (*newVP*). The *Free_Viberation* function is coded as follows:

```
% Update particles based on the free vibration rules.
function newVP=Free_Viberation(VP,HP,GP,BP,D,w1,w2,w3,p)
nVP=size(VP,1);
nV=size(VP,2);

for i=1:nVP
    if p<rand
        w3=0;
        w2=1-w1;
    end
    A=w1*(HP-VP(i,:))+w2*(GP(i,:)-VP(i,:))+w3*(BP(i,:)-VP(i,:));

newVP(i,:)=w1*(D.*A.*rand(1,nV)+HP)+w2*(D.*A.*rand(1,nV)+GP(i,:))+w3*(D.*A.
*rand(1,nV)+BP(i,:));
    w2=0.3;w3=1-w1-w2;
end
```

The fourth function is the regenerating function of the vibrating particles exited from the search space using the harmony search-based handling approach named as *Harmony*. Input arguments are the updated particles (*newVP*), memorized particles (*VP_M*), the parameters of the harmony search-based handling approach (*vpmcr*, *par*, *bw*), and the lower (*Lb*) and the upper (*Ub*) bounds of design variables. The output argument is the corrected set of particles swerving the side limits. It should be noted that the recommended values for the Harmony search-based approach are $vpmcr = 0.95$; $par = 0.1$; and $bw = 0.1$. The *Harmony* function is coded as follows:

```
% If a vibrating particle swerves the side limits correct its position
using the Harmony search based handling approach.

function [newVP]=Harmony(newVP,VP_M,vpmcr,par,bw,Lb,Ub)

for i=1:size(newVP,1)
    for j=1:size(newVP,2)
        if (newVP(i,j)<Lb(j)) || (newVP(i,j)>Ub(j))
            if rand<vpmcr
                newVP(i,j)=VP_M(ceil(rand*size(newVP,1)),j);
                if rand<par
                    if rand<.5
                        newVP(i,j)=newVP(i,j)+bw*rand;
                    else
                        newVP(i,j)=newVP(i,j)-bw*rand;
                    end
                end
            else
                newVP(i,j)=Lb(j)+(Ub(j)-Lb(j))*rand;
            end
        end
    end
end
```

The fifth function is the *Memory* function by which the vibrating particles memory matrix (*VP_M*) and its corresponding vectors of the objective function memory (*Fit_M*) and the penalized objective function memory (*PFit_M*) are updated. The input arguments are *VP*, *Fit*, *PFit*, *VP_M*, *Fit_M*, and *PFit_M*, and the output arguments are *VP_M*, *Fit_M*, and *PFit_M*. The *Memory* function is coded as follows. It should be noted that in the initialization phase, the memorizing should take place firstly. However, the *Memory* function will not be called for memorizing in the initialization phase because of its simplicity in this phase.

```
% Update the vibrating particles memory matrix(VP_M) and its corresponding
vectors of objective function memory (Fit_M) and penalized objective
function memory (PFit_M).

function [VP_M,Fit_M,PFit_M]=Memory(VP,Fit,PFit,VP_M,Fit_M,PFit_M)
nVP=size(VP,1);

for i=1:nVP
    for j=1:nVP
        if PFit(i)<PFit_M(j)
            VP_M(j,:)=VP(i,:);
            Fit_M(j)=Fit(i);
            PFit_M(j)=PFit(i);
            break
        end
    end
end
```

The VPS algorithm coded in the following is composed of three parts: initialization, algorithm cyclic body, and monitoring the results.

```
clc
clear
% rng('default');
%% Initialization

% Define the properties of COP (tension/compression spring design problem).
nV=3; % Number of design variables.
Lb=[0.05 0.25 2]; % Lower bounds of design variables.
Ub=[2 1.3 15]; % Upper bounds of design variables.

% Define the parameters of the VPS algorithm.
nVP=20; % Number of Viberating Particles.
maxNFEs=20000; % Maximum Number of Objective Function Evaluations.
alpha=0.05; % Parameter used in Eq. (13-3)
w1=0.3;w2=0.3;w3=1-w1-w2; % Parameters used in Eq. (13-4)
p=0.1; % With the probability of (1-p) the effect of BP will be ignored in
updating VPs.
vpmcr=0.95;par=0.1;bw=0.1; % Parameters for regenerating the VPs exited
from the search space using the harmony search based handling approach.

% Randomly generate nVP number of vibrating particles as the initial
population of the algorithm.
for i=1:nVP
    VP(i,:)=Lb+(Ub-Lb).*rand(1,nV); % Viberating particles matrix or matrix
of the initial candidate solutions or the initial population..
end

% Evaluate initial population (VP) calling the fobj function constructed in
the second chapter and form its corresponding vectors of objective function
(Fit) and penalized objective function (PFit). It should be noted that the
design vectors all are inside the search space.
for i=1:nVP
    [X,fit,pfit]=fobj(VP(i,:),Lb,Ub);
    VP(i,:)=X;
    Fit(i)=fit;
    PFit(i)=pfit;
end


% Form the viberating particles memory matrix(VP_M) and its corresponding
vectors of objective function memory (Fit_M) and penalized objective
function memory (PFit_M). It should be noted that in the initialization
phase it is not needed to utilize the Memory function and the memory simply
can be produced.
[value,index]=sort(PFit);
VP_M=VP(index,:);Fit_M=Fit(index);PFit_M=PFit(index);
HP=VP_M(1,:); % The best particle obtained sofar by the algorithm.

%% Algorithm Body

NFEs=0; % Current Number of Objective Function Evaluations used by the
algorithm until yet.
NITs=0; % Number of algorithm iterations
maxNITs=maxNFEs/nVP; % Maximum Number of algorithm iterations

while NFEs<maxNFEs
    NITs=NITs+1; % Update the number of algorithm iterations.

    D=(NITs/maxNITs)^(-alpha); % Eq. (13-3)

    % Determine the good (GP) and bad particles (BP) matrixes for updating
the population.
    [GP,BP]=Particles(VP,PFit);
```

```
    % Update particles based on the free vibration rules.
    newVP=Free_Viberation(VP,HP,GP,BP,D,w1,w2,w3,p);

    % Regenerate the particles exited from the search space using the
harmony search based handling approach.
    [newVP]=Harmony(newVP,VP_M,vpmcr,par,bw,Lb,Ub);

    % Evaluate the new vibrating particles. It should be noted that in the
VPS algorithm the replacement strategy is not used. It should be noted also
that the exited dimensions are corrected before within the Harmony
function. However, to do not change the form of fobj function it is checked
again here which is not needed.
    for i=1:nVP
        [X,fit,pfit]=fobj(newVP(i,:),Lb,Ub);
        VP(i,:)=X;
        Fit(i)=fit;
        PFit(i)=pfit;
    end

    % Update the number of Objective Function Evaluations used by the
algorithm until yet.
    NFEs=NFEs+nVP;

    % Update the vibrating particles memory matrix(VP_M) and its
corresponding vectors of objective function memory (Fit_M) and penalized
objective function memory (PFit_M) utilizing the Memory function.
    [VP_M,Fit_M,PFit_M]=Memory(VP,Fit,PFit,VP_M,Fit_M,PFit_M);

    % Monitor the best memorized candidate solution (HP) and its
corresponding objective function (minFit) and penalized objective function
(minPFit).
    MinPFit=PFit_M(1);MinFit=Fit_M(1);HP=VP_M(1,:);

    % Display desired information of the iteration.
    disp(['NITs= ' num2str(NITs) '; MinFit = ' num2str(MinFit) '; MinPFit
= ' num2str(MinPFit)]);

    % Save the required results for post processing and visualization of
the algorithm performance.
    output1(NITs,:)=[MinFit,MinPFit,NFEs];
    output2(NITs,:)=[min(PFit),max(PFit),mean(PFit)];
    output3(NITs,:)=[HP,NFEs];
end

%% Monitoring the results
figure;
plot((1:1:NITs),output2(:,1),'g',(1:1:NITs),output2(:,2),'r--
',(1:1:NITs),output2(:,3),'b-.')
legend('min','max','mean');
xlabel('NITs');
ylabel('PFit');
```

13.4 Experimental Evaluation

Kaveh and Ilchi Ghazaan [2] conducted a deep parametric and sensitivity analysis of VPS to study its search behavior using a benchmark truss problem. The readers are referred to this work for more detail; however, their findings are listed here, considering the number of vibrating particles (*nVP*) equals to 20 results in a better performance of the algorithm in both aspects of accuracy and computational cost and considering α and p equal to 0.05 and 70% results in the better performance of the VPS. Sensitivity analysis on ω_1 and ω_2 indicates that the most suitable performance of the VPS is obtained when the value of 0.3 is considered for these parameters. It should be noted that the parameter ω_3 is not needed to be studied for that obtained as $\omega_3 = 1 - (\omega_1 + \omega_2)$.

The search behavior of the VPS differently conducted and monitored here for two of the main parameters (*nVP* and *p*) of the algorithm for further clarification of its properties. To investigate the effect of the number of vibrating particles, convergence histories for the best and worst and the average of the penalized objective function of the population are depicted in Fig. 13.3 for a single run of the algorithm considering different values of *nVP* (5, 10, 20, and 50). Other parameters are the same as the recommended values listed in the previous paragraph. The convergence histories for the best result of all runs are also depicted in Fig. 13.4. As evidence from these figures, considering 20 numbers of particles is essential to reach a balance between global and local searches.

Such convergence histories are also monitored for the VPS considering different values (0.1, 0.5, 0.7, and 0.9) of the *p* parameter, Figs. 13.5 and 13.6. All runs are simply made using a fixed random series, the same population, and other parameters

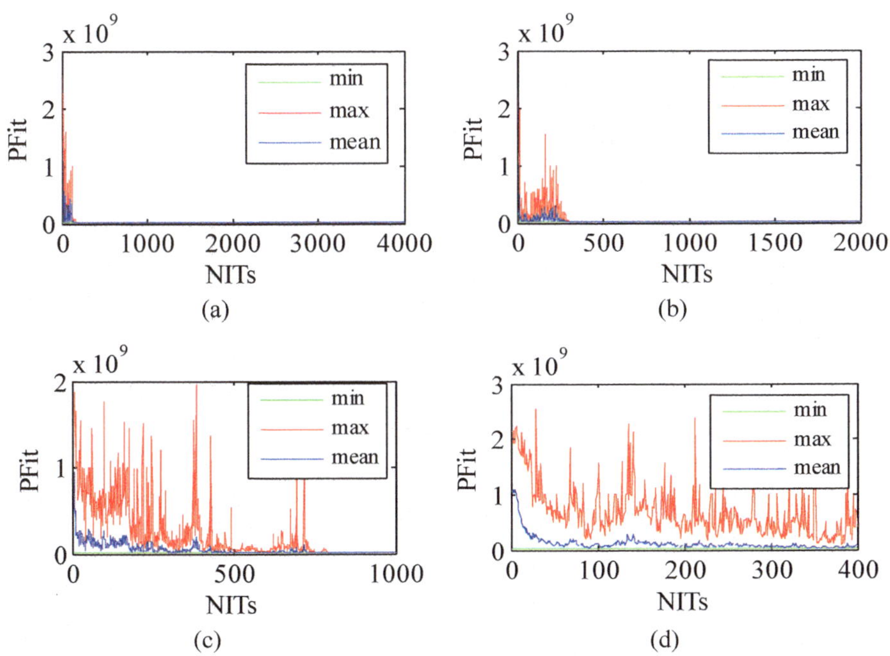

Fig. 13.3 Convergence histories of the minimum, maximum, and mean of the algorithm population considering different values for *nVP*: (**a**) 5; (**b**) 10; (**c**) 20; (**d**) 50

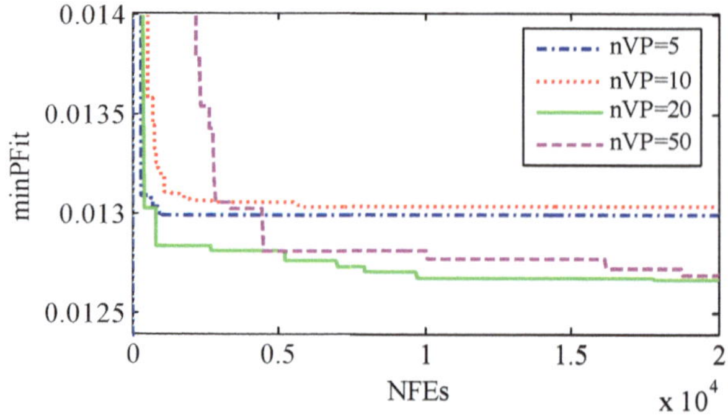

Fig. 13.4 Convergence histories of the VPS considering different values for *nVP*

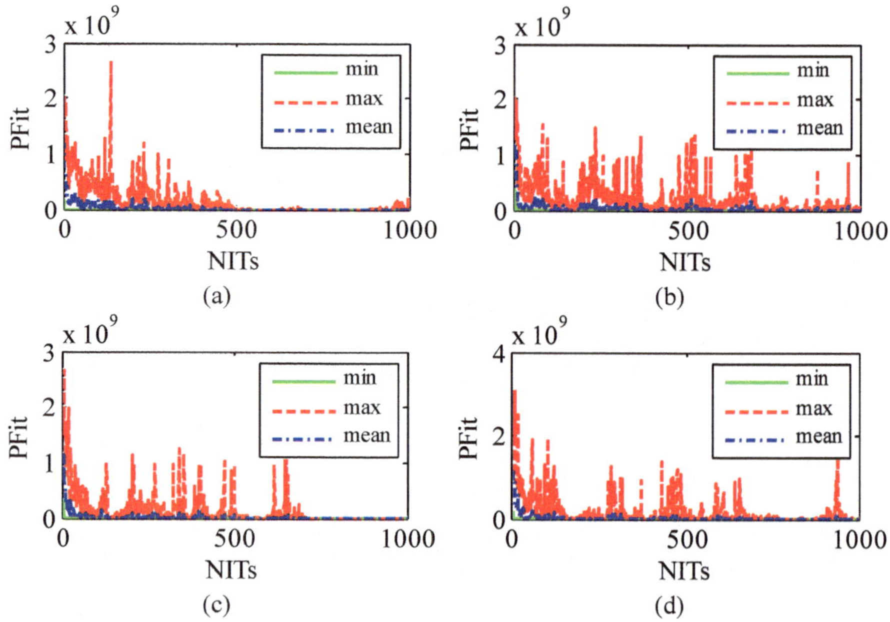

Fig. 13.5 Convergence histories of the minimum, maximum, and mean of the algorithm population considering different values for p: **(a)** 0.1; **(b)** 0.5; **(c)** 0.7; **(d)** 0.9

equal to the ones listed at the first paragraph. It should be remembered that with considering p parameter, VPS ignores the contribution of the bad particles (*BP*) in determining the new equilibrium position of the particles with the probability of $1 - p$. As it is clear, considering large enough value (0.7) is essential. In other words contribution of the bad particles should not be significant in determining the new equilibrium position of the particles.

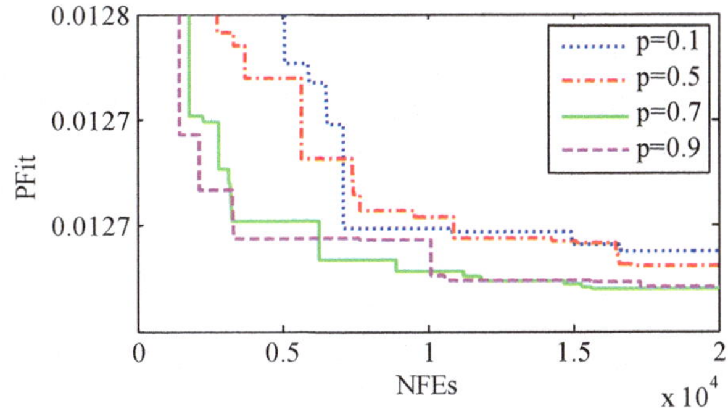

Fig. 13.6 Convergence histories of the VPS considering different values for *nVP*

References

1. Kaveh A, Ilchi Ghazaan M (2017) A new meta-heuristic algorithm: vibrating particles system. Sci Iran Trans A Civ Eng 24(2):551–566
2. Kaveh A, Ilchi Ghazaan M (2017) Vibrating particles system algorithm for truss optimization with multiple natural frequency constraints. Acta Mech 228:307–322
3. Kaveh A, Vazirinia Y (2017) Tower cranes and supply points locating problem using CBO, ECBO, and VPS. Int J Optim Civ Eng 7:393–411
4. Kaveh A, Jafarpour Laien D (2017) Optimal design of reinforced concrete cantilever retaining walls using CBO, ECBO and VPS algorithms. Asian J Civ Eng 18:657–671
5. Shahrouzi M, Farah-Abadi H (2018) Optimal seismic design of steel moment frames by un-damped multi-objective vibrating particles system. Asian J Civ Eng 19:877–891

Chapter 14
Cyclical Parthenogenesis Algorithm

14.1 Introduction

This chapter presents Cyclical Parthenogenesis Algorithm (CPA) proposed by Kaveh and Zolghadr [1]. CPA is inspired by reproduction and social behavior of some zoological species like aphids, which can reproduce with and without mating. It is applied by Kaveh and Zolghadr [2–4] successfully for some structural optimization problems.

Like other population-based metaheuristics, CPA starts with a set of random initial candidate solutions. The algorithm considers each candidate solution as a living organism like aphid and groups them into a number of colonies with the same number of aphids each inhabiting a host plant. Each colony tries to iteratively improve the quality of its aphids by reproduction mechanisms with and without mating and by a chance to get merit from other colonies using information exchange mechanism. In each colony, the best aphids are considered as female and reproduce without mating by which the offspring arise from the female parent and inherit the genes of that parent only. The worse ones are considered as males and reproduce with mating by which offspring arise from the male parent and one of the randomly selected females. Aphids have another intricate ability: producing winged offspring in response to poor conditions on the host plant. Inspiring this ability, CPA uses an information exchange mechanism by which in each iteration, with a certain probability, two colonies are selected randomly and the best aphid of one of the colonies replaced with the worst one of the other colony. Colony improvements and information exchange between them are repeated in the cyclic body of the algorithm in succession to satisfy stopping criteria with the aim of directing each colony toward the favorable position in the search space.

A. Kaveh, T. Bakhshpoori, *Metaheuristics: Outlines, MATLAB Codes and Examples*, https://doi.org/10.1007/978-3-030-04067-3_14

14.2 Formulation and Framework of the Cyclical Parthenogenesis Algorithm

Aphids are one of the highly successful organisms of the superfamily Aphidoidea. In the following firstly some key aspects of intricate cyclical parthenogenesis of aphids as the inspiration basis for the main mechanisms of CPA are presented. Afterward, the concepts and framework of CPA algorithm, its pseudo code, and the related flowchart are presented.

14.2.1 Aphids and Cyclical Parthenogenesis

Aphids are small sap-sucking insects and members of the superfamily Aphidoidea. Aphids are known as one of the most destructive insect pests on cultivated plants in temperate regions. This is mainly because of their intricate life cycles and close association with their host plants. Some common features of this intricate life cycle are found to be interesting by Kaveh and Zolghadr [1] from an optimization point of view.

Aphids are capable of reproducing offspring with and without mating. When reproducing without mating, the offspring arise only from the female parent and inherit the genes of that parent only. In this type of reproduction, most of the offspring are genetically identical to their mother, and genetic changes occur rarely. This form of reproduction is chosen by female aphids in suitable and stable environments and allows them to rapidly grow a population of similar aphids, which can exploit the favorable circumstances. Reproduction through mating, on the other hand, offers a net advantage by allowing more rapid generation of genetic diversity, making adaptation to changing environments available.

Some aphid species produce winged offspring in response to poor conditions on the host plant or when the population on the plant becomes too large. These winged offspring can disperse to other food sources. Flying aphids have little control over the direction of their flight because of their low speed. However, once within the layer of relatively still air around vegetation, aphids can control their landing on plants and respond to olfactory or visual cues or both.

Although the reasons behind reproduction and social behavior of aphids are not completely agreed upon in zoology, advantages of these aspects are evident from an optimization point of view.

14.2.2 CPA Algorithm

Looking at the key aspects of cyclical parthenogenesis of aphids overviewed at the previous subsection, there is an analogy between it and a population-based metaheuristic. Each agent of the algorithm can be considered as an aphid. All the aphids can be divided into some number of colonies. The role (female or male) of each aphid in each colony can be determined based on the quality. Each colony reproduces independently with the aim of improving the position of its aphids in the search space or reaching to a favorable circumstances with very high rates of reproduction. On the other hand, to prevent the reproduction of colonies in an independent manner, benefiting the winged aphids, colonies can exchange a level of information between themselves. The rules of CPA are stated at the following:

Rule 1: Initialization

CPA starts from a set of candidate solutions or aphids randomly generated within the search space. The number of aphids is considered as *nA*. These aphids are grouped into *nC* number of colonies with the same number of members or aphids (*nM*). The concept of multiple colonies allows CPA to search different portions of the search space more or less independently and prevents the unwanted premature convergence phenomenon. For coding CPA in a simple manner and to be easy for tracing, the colonized aphids are determined by a cell array (*CA*). Therefore, *CA* is an array of *nC* colonies with *nM* aphids. After evaluation of the initial population or the colonized aphids, the corresponding objective function (*Fit*) and penalized objective function (*PFit*) cells are produced.

According to this rule, *nM* is not considered as a population parameter of the algorithm, so that can be calculated using two population parameters of the algorithm: $nM = nA/nC$. It should be noted that CPA considers *nM* unchanged in the optimization procedure.

Rule 2: Reproduction or Parthenogenesis of Aphids

In each iteration, *nM* new candidate solutions or offspring are generated in each of the colonies. These new solutions can be reproduced either with or without mating. A ratio *Fr* of the best of the new solutions of any colony are considered as female aphids; the rest are considered as male aphids. Therefore in each colony, $Fr \times nM$ number of offspring will be reproduced without mating, and $(1 - Fr) \times nM$ number of offspring will be reproduced with mating. Altogether *nM* number of offspring will be reproduced.

For reproducing $Fr \times nM$ number of offspring without mating, a female parent (*F*) is selected randomly from the female aphids of the colony for $Fr \times nM$ times. Then, this randomly selected female parent reproduces a new offspring without mating by the following expression:

$$newCA = F + \alpha_1 \times \frac{randn}{NITs} \times (Ub - Lb) \tag{14.1}$$

where *randn* is a random number drawn from a normal distribution, *NITs* is the current number of algorithm iteration, and α_1 is a scaling parameter for controlling step size of searching.

In order to reproduce $(1 - Fr) \times nM$ number of offspring, each of the male aphids (M) selects a female aphid (F) randomly in order to produce an offspring through mating:

$$newCA = M + \alpha_2 \times rand \times (F - M) \tag{14.2}$$

where *rand* is a random number uniformly distributed within (0,1) interval and α_2 is a scaling parameter for controlling searching step size. It can be seen that in this type of reproduction, two different solutions share information, while when reproduction occurs without mating, the new solution is generated using merely the information of one single parent solution.

Rule 3: Death and Flight
When all of the new solutions or offspring of all colonies are generated and the objective function values are evaluated, flying occurs with a probability of *Pf* where two of the colonies are selected randomly and named as *colony*1 and *colony*2. A winged aphid is reproduced by and identical to the best female of *colony*1 and then flies to *colony*2. In order to keep the number of members of each colony constant, it is assumed that the worst member of *colony*2 dies.

Parameter *Pf* is responsible for defining the level of information exchange among the colonies. With no possible flights ($Pf = 0$), the colonies would be performing their search in a completely independent manner, i.e., an optimization runs with *nA* aphids divided into *nC* colonies would be similar to *nC*-independent runs each with $nM = nA/nC$ aphids in one colony. It is obvious that this would not be particularly favorable since it is, in fact, changing the population of aphids without actually utilizing the abovementioned benefits of the multiple colonies. On the other hand, permitting too many flights ($Pf = 1$) results in the same effect by merging the information sources of different colonies. It is important to note that at the early stages of the optimization process, it is more favorable to give the colonies a higher level of independence so that they can search the problem space without being affected by the other colonies. However, as the optimization process proceeds, it is desirable to let the colonies share more information so as to provide the opportunity for the more promising regions of the search space to be searched thoroughly. Considering *Pf* linearly increasing from 0 to 1 results in the best performance of the algorithm, since it conforms to the abovementioned discussion on information circulation:

$$Pf = (NITs - 1)/(maxNITs - 1) \tag{14.3}$$

Rule 4: Updating the Colonies or the Replacement Strategy
Considering the fact that the aphids of each colony are capable of reproducing a genetically identical offspring without mating, CPA compares the newly generated set of offspring based on Rule 2 for each colony with the current position of the colony and transmits the *nM* best ones for the next iteration.

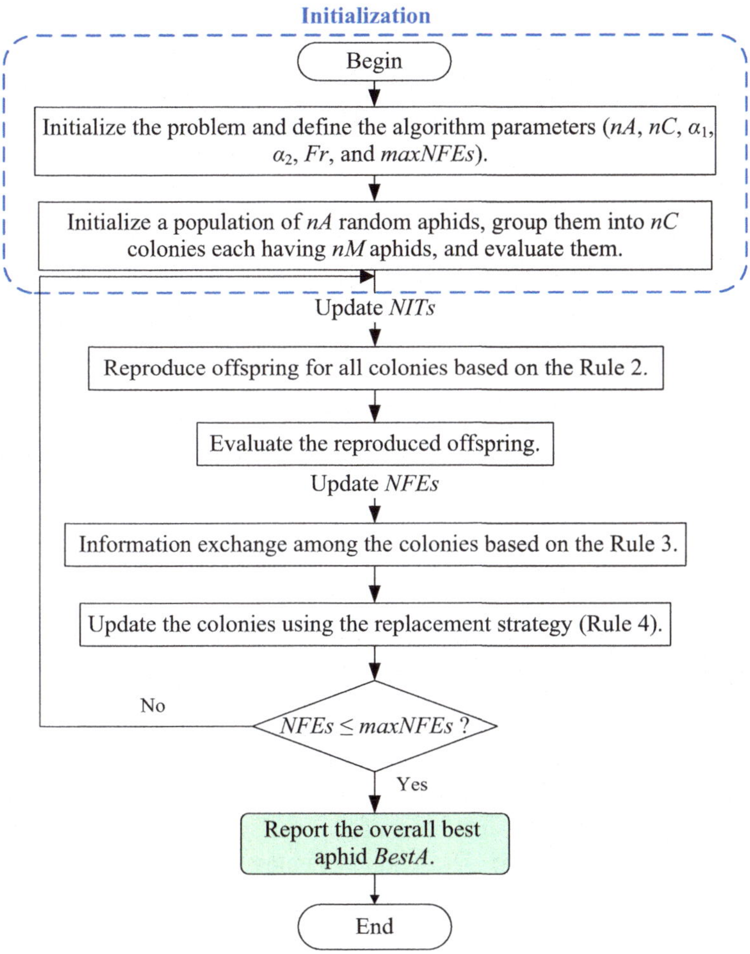

Fig. 14.1 Flowchart of the CPA algorithm

Rule 5: Termination Criteria

A maximum number of objective function evaluations (*maxNFEs*) or a maximum number of algorithm iterations (*maxNITs*) is considered as the stopping criterion.

The pseudo code of algorithm is provided as follows, and the flowchart of CPA is illustrated in Fig. 14.1.

The pseudo code of CPA algorithm for solving COPs:

Define the algorithm parameters: *nA*, *nC*, α_1, α_2, *Fr*, and *maxNFEs*.

Initialize a population of *nA* random aphids, and group them into *nC* colonies with each having *nM* aphids.

Evaluate initial population (*CA*), and form its corresponding vectors of objective function (*Fit*) and penalized objective function (*PFit*) cells.

While *NFEs* < *maxNFEs*

Update the number of algorithm iterations (*NITs*).
Reproduce offspring for all colonies without and with matting by dividing each colony into female and male considering *Fr* (Rule 2).
Evaluate the reproduced offspring.
Update *NFEs*.
Information exchange among the colonies with the probability of *Pf* by flying a winged aphid from a randomly selected colony to another one (Rule 3).
Update the colonies using the replacement strategy (Rule 4).
Monitor the best aphid (*bestA*) of each colony and the overall best aphid (*BestA*) of all colonies.

end While

14.3 MATLAB Code for Cyclical Parthenogenesis Algorithm (CPA)

It should be noted that to code the algorithm in a simple manner so that the reader can trace the code line by line and even easily realize all the items defined or generated in MATLAB environment, using ready functions and structure arrays is avoided. Instead of the structure arrays, the cell arrays are benefited. According to the previous section, four functions are considered for coding the algorithm. First, these functions are presented with the order of their recall in the algorithm, and at the end, the algorithm is coded. The required comments are presented. Note that the percent sign (%) represents a comment.

The first function is the objective function evaluation (*fobj*) which was presented in Chap. 2.

The second function is named as the *Parthenogenesis* function for reproducing with and without mating. The input arguments are the current set of colonized aphids (*CA*) and its corresponding penalized objective function cell (*PFit*), the lower and upper bound of design variables (*Lb* and *Ub*), search step size controlling parameters (α_1 and α_2), *Fr* parameter to determine the ratio of aphids of each colony to be considered as a female, and the current number of the algorithm iterations (*NITs*). The only output argument is the reproduced set of offspring (*newCA*). It should be noted that to determine the female and male aphids in each colony, another cell named as parents (*P*) should be generated simply by sorting the aphids of each colony according to their penalized objective function. In each parent cell of colonies, the *Fr* ratio of sorted aphids are females, and the remaining ones are males. The *Parthenogenesis* function is coded as follows:

```
% Reproducing with and without mating.

function [newCA]=Parthenogenesis(CA,PFit,Lb,Ub,Fr,alpha1,alpha2,NITs)
nC=size(CA,2);
nM=size(CA{1},1);
nV=size(CA{1},2);

% Determine the parents (females or males) in each colony. This can be made
simply by sorting according to the penalized objective function. In each
colony the Fr ratio of sorted aphids are females and the remaining ones are
males.
for i=1:nC
    [~,index]=sort(PFit{i});
    P{i}=CA{i}(index,:);
end

for i=1:nC
    for j=1:nM
        rfi=round(1+(Fr*nM-1).*rand); % The index of random female parent
for reproduction with and without mating.
        F=P{i}(rfi,:); % Randomly selected female.
        if j<=Fr*nM
            newCA{i}(j,:)=F+alpha1/NITs*randn(1,nV).*(Ub-Lb); % New aphid
generated without mating.
        else
            M=CA{i}(j,:); % The male aphid.
            newCA{i}(j,:)=M+alpha2*rand(1,nV).*(F-M); % New aphid generated
with mating.
        end

    end
end
```

The third function is considered for information exchange between colonies based on Rule 3 and named as *Flying* function. The input arguments are the newly reproduced offspring (*newCA*) and its corresponding objective (*newFit*) and penalize objective function (*newPFit*) cells. Information exchange or flying is accrued by a linearly increasing probability function (Eq. 14.3). Therefore, the current number of algorithm iterations (*NITs*) and the maximum number of algorithm iterations (*maxNITs*) are the other input arguments of this function. Output arguments are the same as the first three input arguments with a level of exchanged information.

```
% Information exchange among the colonies by flying a winged aphid from a
randomly selected colony to another one.

function [newCA,newFit,newPFit]=Flying(newCA,newFit,newPFit,NITs,maxNITs)

nC=size(newCA,2);

Pf=(NITs-1)/(maxNITs-1); % Probability of flying as a linearly increasing
function from 0 to 1.

if rand<Pf
    i1=round(1+(nC-1).*rand); % Index of the first randomly selected
colony.
    i2=i1;
    while i2==i1
        i2=round(1+(nC-1).*rand); % Index of the second randomly selected
colony.
    end
    [~,i3]=min(newPFit{i1}); % Index of the winged aphid which is the best
aphid of the first randomly selected colony.
    [~,i4]=max(newPFit{i2}); % Index of the dead aphid which is the worst
aphid of the second randomly selected colony.

    % Replace the dead aphid with the winged one.
    newCA{i2}(i4,:)=newCA{i1}(i3,:);
    newFit{i2}(i4)=newFit{i1}(i3);
    newPFit{i2}(i4)=newPFit{i1}(i3);
end
```

The last function is the *Replacement* function in which the colonized population will be updated. The current set of colonized aphids and the newly generated offspring compared according to the penalized objective function and the better ones are transmitted to the next iteration. The *Replacement* function is coded as follows:

```
% Update the colonies by comparing the reproduced ones.

function [CA,Fit,PFit]=Replacement(CA,Fit,PFit,newCA,newFit,newPFit)
nC=size(newCA,2);
nM=size(newCA{1},1);

for i=1:nC
    helpCA=[CA{i};newCA{i}];helpFit=[Fit{i}     newFit{i}];helpPFit=[PFit{i}
newPFit{i}];
    [~,order]=sort(helpPFit);
    CA{i}=helpCA(order(1:nM),:);
    Fit{i}=helpFit(order(1:nM));
    PFit{i}=helpPFit(order(1:nM));
end
```

CPA algorithm is coded in the following which is composed of three parts: initialization, algorithm cyclic body, and monitoring the results.

```
clc
clear
%% Initialization

% Define the properties of COP (tension/compression spring design problem).
nV=3; % Number of design variables.
Lb=[0.05 0.25 2]; % Lower bounds of design variables.
Ub=[2 1.3 14]; % Upper bounds of design variables.

% Define parameters of CPA algorithm.
nA=60; % Number of Aphids
nC=4; % Number of Colonies
nM=nA/nC; % Number of aphids in each colony which is equal and unchanged.
alpha1=1;alpha2=2; % Search step size parameters.
Fr=0.4; % Parameter to determine the ratio of aphids of each colony to be
considered as female.
maxNFEs=20000; % Maximum number of Objective Function Evaluations.

% Initialize a population of nA random candidate solutions and group them
into nC colonies with each having nM members or aphids. For simplicity a
cell array is used.
for i=1:nC
    for j=1:nM
        CA{i}(j,:)=Lb+(Ub-Lb).*rand(1,nV); % Colonized aphids.
    end
end

%Evaluate grouped initial population or colonized aphids (CA) calling the
fobj function constructed in the second chapter and form its corresponding
cells of objective function (Fit) and penalized objective function (PFit).
It should be noted that the design vectors all are inside the search space.
for i=1:nC
    for j=1:nM
        [X,fit,pfit]=fobj(CA{i}(j,:),Lb,Ub);
        CA{i}(j,:)=X;
        Fit{i}(j)=fit;
        PFit{i}(j)=pfit;
    end
end

%% Algorithm Body

NFEs=0; % Current number of Objective Function Evaluations used by the
algorithm until yet.
NITs=0; % Number of algorithm iterations
maxNITs=round(maxNFEs/nA); % Maximum number of algorithm iterations.

while NFEs<maxNFEs
    NITs=NITs+1; % Update the number of algorithm iterations.

    % Reproducing with and without mating.
    [newCA]=Parthenogenesis(CA,PFit,Lb,Ub,Fr,alpha1,alpha2,NITs);

    % Evaluate the reproduced aphids calling the fobj function.
    for i=1:nC
        for j=1:nM
            [X,fit,pfit]=fobj(newCA{i}(j,:),Lb,Ub);
            newCA{i}(j,:)=X;
            newFit{i}(j)=fit;
            newPFit{i}(j)=pfit;
        end
    end
    NFEs=NFEs+nA;
```

```
    %Information exchange among the colonies by flying a winged aphid from
a randomly selected colony to another one.
    [newCA,newFit,newPFit]=Flying(newCA,newFit,newPFit,NITs,maxNITs);

    % Update the colonies by the replacement strategy.
    [CA,Fit,PFit]=Replacement(CA,Fit,PFit,newCA,newFit,newPFit);

    % Monitor the best aphid (bestA) and its corresponding penalized
objective function (minPFit) and objective function (minFit) for each
colony.
    for i=1:nC
        [minPFit(i),m]=min(PFit{i});
        minFit(i)=Fit{i}(m);
        bestA(i,:)=CA{i}(m,:);
    end
    % Monitor the overall best aphid (BestA) and its corresponding
penalized objective function (MinPFit) and objective function (MinFit).
    [MinPFit,m]=min(minPFit);
    MinFit=minFit(m);
    BestA=bestA(m,:);

    % Display desired information of the iteration.
    disp(['NITs= '  num2str(NITs) '; MinFit = '  num2str(MinFit) '; MinPFit
= '  num2str(MinPFit)]);

    %Save the required results for post processing and visualization of
algorithm performance.
    output1(NITs,:)=[MinFit,MinPFit,NFEs];
    output2(NITs,:)=[MinFit,max(cell2mat(PFit)),mean(cell2mat(PFit))];
    output3(NITs,:)=[BestA,NFEs];
end

%% Monitoring the results
figure;
plot((1:1:NITs),output2(:,1),'g',(1:1:NITs),output2(:,2),'r--
',(1:1:NITs),output2(:,3),'b-.')
legend('min','max','mean');
xlabel('NITs');
ylabel('PFit');
```

14.4 Experimental Evaluation

CPA has six parameters: number of aphids (*nA*), number of colonies (*nC*), parameters to control the searching step sizes (α_1 and α_2), parameter to determine the ratio of aphids of each colony to be considered as female (*Fr*), and the maximum number of objective function evaluations as the stopping criteria (*maxNFEs*).

Kaveh and Zolghadr [2] conducted a deep parametric and sensitivity analysis of CPA algorithm to study its search behavior using a benchmark truss problem. The readers are referred to this work for more detail; however, their findings are listed here.

nA: Considering at least 60 numbers of aphids results in better performance of the algorithm.

nC: This parameter value provides a good balance between the diversification and intensification tendencies of the algorithm. In fact, dividing the population of aphids into more colonies limits the circulation of information among the aphids and allows the aphids of a colony to explore the search space more independently without being affected by other colonies. This generally results in a more diverse search. On the other hand, having more aphids in a colony (less number of colonies) permits the algorithm to search a particular region of the search space more thoroughly (more intensification) but, at the same time, limits the diversification of the algorithm. $nC = 4$ provides a good balance between the exploration and exploitation tendencies of the algorithm.

Fr: This parameter defines the ratio of aphids of each colony to be considered as a female. Increasing the value of this parameter results in an increase in the number of aphids generated without mating in the subsequent generation. Since such aphids use a single source of information (female parent), they can be interpreted as a means of searching for a localized region around their parent. This localized region gets smaller gradually as the optimization process proceeds. Reproduction through mating, on the other hand, incorporates two different sources of information and consequently contributes to the convergence of the algorithm by letting the aphids share information. CPA exhibits its best performance when the value of this parameter is taken as 0.4.

α_1 and α_2: These parameters represent the step size of the aphids in reproduction without and with mating, respectively. The best performance of the algorithm corresponds to $\alpha_1 = 1$ and $\alpha_2 = 2$.

References

1. Kaveh A, Zolghadr A (2017) Cyclical parthenogenesis algorithm: a new meta-heuristic algorithm. Asian J Civil Eng 18:673–701
2. Kaveh A, Zolghadr A (2017) Cyclical parthenogenesis algorithm for guided modal strain energy based structural damage detection. Appl Soft Comput 57:250–264
3. Kaveh A, Zolghadr A (2018) Optimal design of cyclically symmetric trusses with frequency constraints using cyclical parthenogenesis algorithm. Adv Struct Eng 21:739–755
4. Kaveh A, Zolghadr A (2018) Meta-heuristic methods for optimization of truss structures with vibration frequency constraints. Acta Mech 229(October):1–22

Chapter 15
Thermal Exchange Optimization Algorithm

15.1 Introduction

Based on Newton's law of cooling, Kaveh and Dadras [1] introduced a new metaheuristic named Thermal Exchange Optimization (TEO) algorithm. TEO is in early stages of its use and is utilized to solve a few structural optimization problems [2, 3]. Newton's law of cooling states that the rate of heat loss of a body is proportional to the difference in temperatures between the body and its surroundings. TEO considers each of its particles as a cooling or heating object, and by associating another agent as the environment, a heat transferring and thermal exchange happens between them. The new temperature of the object is considered as its next position in the search space.

Like other metaheuristics, TEO starts from a set of randomly generated initial candidate solutions. In each iteration of the algorithm, all agents of the population are evaluated and sorted according to their objective function values. Then the population is divided into two parts with the same number of objects. All objects will be affected by the environmental temperature, no matter which group they belong. If the object belongs to the first half, its environmental temperature will be its corresponding object from the second part and vice versa. The best solutions which are the objects from the first half and have more temperature are cooled by moving slightly toward the particles with a lower temperature. The bad particles which are the objects from the second half and have low temperature are heated by moving toward the particles with higher temperature. The heat transferring between the objects takes place in the cyclic body of the algorithm with the aim of conducting all particles to the better positions without any difference in the temperature. It should be mentioned that TEO benefits a memory with a specific size to save the best-so-far known solutions. TEO used these memorized particles for replacing the same number of worst ones in each of its iterations.

A. Kaveh, T. Bakhshpoori, *Metaheuristics: Outlines, MATLAB Codes and Examples*, https://doi.org/10.1007/978-3-030-04067-3_15

15.2 Formulation and Framework of TEO

In the following, after overviewing Newton's law of cooling as the essence of TEO, the concepts and framework of TEO algorithm, its pseudo code, and the related flowchart are presented.

15.2.1 Newton's Law of Cooling

At the first stage, the temperature solution for the body with lumped thermal capacity is presented, by which TEO updates the position of the agents in the search space. Assume that the overall heat transfer coefficient of a body is equal to h (in W m^2 K^{-1}) and the object has high temperature T_0 (in °K) at time $t = 0$ (in sec) and is suddenly placed in a different environment where it is cooled by surrounding fluid at a constant temperature T_b (in °K). The volume of the solid is V (in m^3) and its surface area, which the heat flow (Q) takes place on it, is A (in m^2). The rate of heat loss from the surface is:

$$dQ/dt = h(T_0 - T_b)A \tag{15.1}$$

The heat loss in time dt is $h(T_0 - T_b)Adt$, and this is equal to the change in the stored heat as the temperature falls dT, i.e.:

$$V\rho c dT = -hA(T - T_b)dt \tag{15.2}$$

where ρ and c are the density (kg m^{-3}) and specific heat (J kg^{-1} K^{-1}), respectively. Integration results in:

$$\frac{T - T_b}{T_0 - T_b} = \exp\left(-\frac{hA}{V\rho c}t\right) \tag{15.3}$$

Assuming the coefficients behind the t as β, the result of the integration can be rearranged as follows:

$$T = T_b + (T_0 - T_b)\exp(-\beta t) \tag{15.4}$$

and this is the main physics-based formulation by which TEO is inspired. This law simply states that the temperature of a hot (or cold) object progresses toward the temperature of its environment in a simple exponential fashion and the rate of temperature exchange is proportional to difference between the object and the environment.

15.2.2 TEO Algorithm

Looking at the temperature solution for the body with lumped thermal capacity overviewed at the previous subsection, there is an analogy between it and a population-based metaheuristic. Each agent of the algorithm can be considered as a cooling or heating object (thermal object). By associating again each agent as the environment, a heat transferring and thermal exchange happens between them. The new temperature of the object is considered as its next position in the search space. In the following the TEO algorithm is presented in the form of six rules:

Rule 1: Initialization
TEO starts from a set of candidate solutions randomly generated within the search space. The number of thermal objects is considered as nTO. These particles form the matrix of thermal objects (TO). After evaluating the objects, the corresponding objective function (Fit) and the penalized objective function ($PFit$) vectors are produced.

Rule 2: Creating Groups
For heat transferring process, first, the objects should be sorted in an ascending order based on their penalized objective function and then be categorized into two groups with a same number of objects. In this way, a better solution is known as a warmer object. This type of pairing is originally proposed in CBO algorithm (Chap. 10) and depicted in Fig. 15.1. In this figure for more clarity, the number of thermal objects is considered as n. Any object (hot or cold) progresses toward the temperature of its environment. Therefore, considering the first half ($i = 1, \ldots, nTO/2$) as the cooling objects, the remaining half ($i = nTO/2 + 1, \ldots, nTO$) will rule as their environmental objects and vice versa; considering the second half as the heating objects, the first half will rule as their environmental objects.

Rule 3: Heat Transferring or Updating the Objects
Based on the temperature solution for the body with lumped thermal capacity, Eq. (15.4), each body (hot or cold) will move toward to the corresponding environmental object ($envTO$) as follows:

$$newTO(i) = envTO(i) + (TO(i) - envTO(i))\,\exp(-\beta(i)t) \tag{15.5}$$

in which $newTO(i)$ is the new position of the cooled or heated object. According to this formulation, any object will be regenerated, or updated, or cooled, or heated

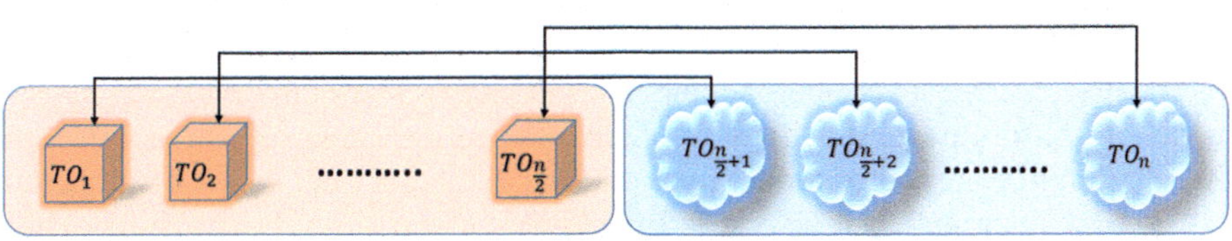

Fig. 15.1 The pairing of objects for temperature progressing

around its corresponding environmental temperature with a step size which is a simple exponential function, and its amplitude is proportional to the difference between object and environment.

TEO considers the time and β parameters heuristically as follows:

Time is associated with the iteration number in a way that the value of t is considered as:

$$t = \frac{NITs}{maxNITs} \tag{15.6}$$

where $NITs$ is the current number of algorithm iterations and $maxNITs$ is the maximum number of algorithm iterations considered as the stopping criteria of the algorithm. According to this equation, t grows from a near-zero value to 1 with the progress of the algorithm. Therefore, the time is targeted in a way that results in a decreasing exponential function step size. Such a function enables TEO to have diversification with larger search step sizes in its earliest iterations and to gradually transient to the more intensification with the smallest search step sizes.

Parameter β for each agent is considered as:

$$\beta(i) = \frac{PFit(i)}{maxPFit} \tag{15.7}$$

where $maxPFit$ is the penalized objective function of the worst or the coldest object in the current population. Note that it is considered as the denominator in the equation. Note also that the value of β is directly proportional to the penalized objective function of each object. Therefore, the best or warmest object will be assigned by the smallest value of β, and the worst or coldest object will be assigned by 1. Such a definition results in larger step sizes for best objects and smallest step sizes for the worst particles. Notice that environmental objects for the best objects are the worst particles and vice versa. Therefore such a definition for the β enables TEO to update the best objects to positions far from the worst ones and update the worst particles to positions near to the best particles. And this reflects the elitism of the algorithm.

Rule 4: Randomness

Defining Eq. (15.5) and parameters used in it, arise in the mind that TEO is a rule-based algorithm. However, metaheuristic algorithms should have the ability to escape from traps, when agents get close to a local optimum. Randomness is an integral part of the metaheuristics to solve this problem. In this regard, TEO benefits two mechanisms to incorporate the randomness in its framework.

The first mechanism is incorporating the randomness in the environmental objects as the origin for updating the corresponding cooled or heated objects. All thermal objects should be modified randomly according to the following mechanism before using as the environmental objects in Eq. (15.5).

$$envTO = (1 - c \times rand)\ TO;\ c = c_1 + c_2(1 - t); \tag{15.8}$$

where c_1 and c_2 are the controlling parameters chosen from {0 or 1}. $(1 - t)$ decreases linearly the randomness and increases exploitation by nearing to the last iterations. c_2 controls $(1 - t)$. For instance, this can be considered equal to zero, when decreasing is not required. c_1 controls the size of the random steps. Furthermore, when a decreasing process is not employed ($c_2 = 0$; as said above), c_1 involves the randomness. In the case of considering both control parameters equal to zero ($c = 0$), none of the abovementioned mechanisms are employed, and any object will be considered as environmental temperature without any modification.

Considering a probability parameter (p) is the second mechanism by which TEO updates a randomly selected component of any cooling or heating object. For each object, p is compared with *rand* which is a random number uniformly distributed within (0, 1). If $rand < p$, one component of the agent is regenerated randomly from the search space.

Rule 5: Considering a Memory to Save the Best Objects

TEO considers a memory which saves some historically best objects (TO_M) and their related objective function (Fit_M) and penalized objective function (PFit_M) values. The size of the memory is considered as *STO_M*. Thermal objects memory can be used for many purposes such as position correction of the objects exited from the search space and guidance of the current objects. As it is obvious from previous rules, TEO does not use the replacement strategy and guaranties the elitism of the algorithm by Rule 3. TEO uses simply the memory to replace the same number of current worst objects with the memorized ones in each iteration, and in this way the elitism of the algorithm will be strengthened.

Rule 6: Termination Criteria

A maximum number of objective function evaluations (*maxNFEs*) or a maximum number of algorithm iterations (*maxNITs*) is considered as the terminating criterion.

The pseudo code of TEO is given as follows, and the flowchart is illustrated in Fig. 15.2.

The pseudo code of TEO algorithm for solving COPs:

Define the algorithm parameters: *nTO*, *p*, *STO_M*, and *maxNFEs*.
Generate random initial thermal objects (*TO*).
Evaluate the initial thermal objects, and form its corresponding vectors of the objective function (*Fit*) and penalized objective function (*PFit*).
Form the thermal objects memory matrix (*TO_M*) and its corresponding vectors of objective function memory (*Fit_M*) and penalized objective function memory (*PFit_M*).
While *NFEs*<*maxNFEs*

  Update the number of algorithm iterations (*NITs*).
  Determine the time step using Eq. (15.6).
  Determine the environmental temperature of the objects based on the Eq. (15.8).
  Update objects based on Eq. (15.5).

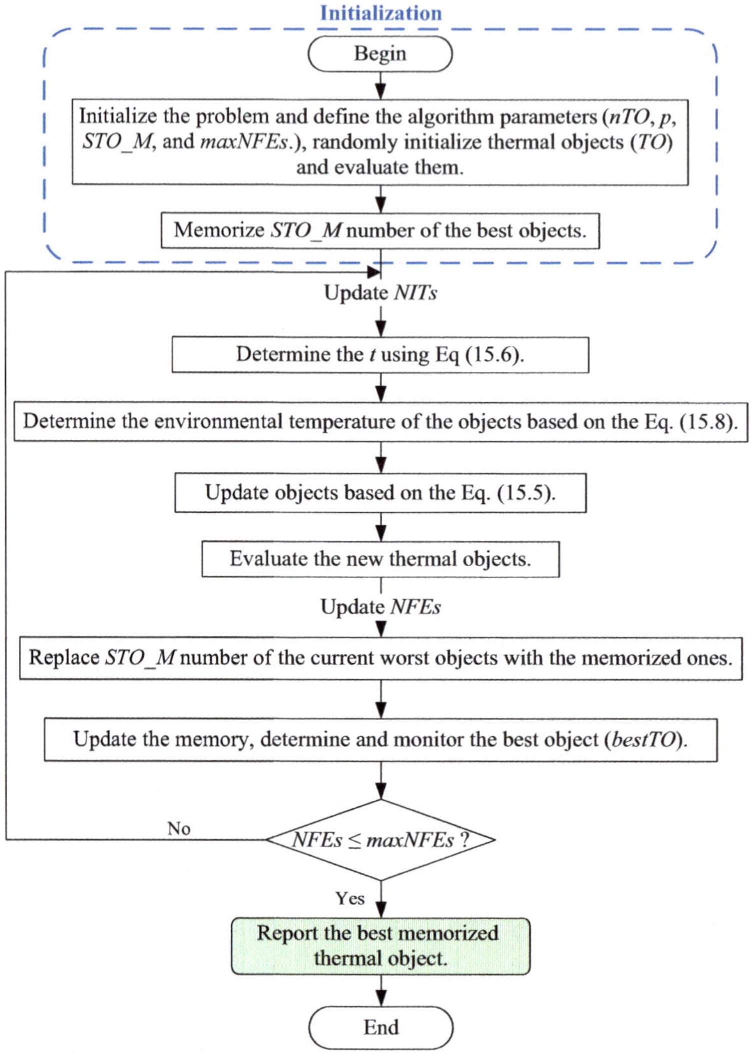

Fig. 15.2 Flowchart of the TEO algorithm

Evaluate the new thermal objects.
Update *NFEs*.
Replace *STO_M* number of the current worst objects with the memorized ones.
Update the thermal objects memory matrix (*TO_M*) and its corresponding vectors of objective function memory (*Fit_M*) and penalized objective function memory (*PFit_M*).
Determine and monitor the best memorized candidate solution (*bestTO*).

end While

15.3 MATLAB Code for the Thermal Exchange Optimization (TEO) Algorithm

According to the previous section, five functions are considered for coding the algorithm. First, these functions are presented with the order of their recall in the algorithm, and at the end, the algorithm is coded. The required comments are presented. Note that the percent sign (%) represents a comment.

The first function is the objective function evaluation (*fobj*) which was presented in Chap. 2.

The second function named as the *Environment* function to determine the environmental temperature of objects (*envTO*) will be used in updating the population. The input arguments are the current position of the thermal objects matrix (*TO*) and time step (*t*). The output argument is the *envTO* matrix.

```
% Determine the environmental temperature of the objects.
function [envTO]=environment(TO,t)

nTO=size(TO,1);
nV=size(TO,2);

for i=1:nTO
    c1=round(rand);c2=round(rand);
    c=c1+c2*(1-t);
    envTO(i,:)=(ones(1,nV)-c*rand(1,nV)).*TO(i,:);
end
```

The third function is the *Newtonslaw* function in which the thermal objects are updated to a new position or temperature. The input arguments are the current set of thermal objects (*TO*), corresponding environmental objects (*envTO*), objective function (*Fit*), and the penalized objective function (*PFit*) matrixes, the probability parameter (*p*), time step (*t*), and the lower (*Lb*) and upper (*Ub*) bound of design variables. The output argument is the updated thermal object matrix (*newTO*). The *Newtonslaw* function is coded in the follows:

```
% Temperature exchange of objects.
function [newTO]=Newtonslaw(TO,envTO,Fit,PFit,p,t,Lb,Ub)

nTO=size(TO,1);
nV=size(TO,2);

% It should be noted that sorting also will be made in the Replacement
function. Therefore, it is needed here only in the first iteration.
[~,order]=sort(PFit);
TO=TO(order,:);
Fit=Fit(order);
PFit=PFit(order);
envTO=envTO(order,:);

for i=1:nTO
    beta=PFit(i)/PFit(nTO);
    if i<=nTO/2
        newTO(i,:)=envTO(nTO/2+i,:)+(TO(i,:)-envTO(nTO/2+i,:))*exp(-
beta*t);
        if rand<p
            index=ceil(rand*nV);
            newTO(i,index)=Lb(index)+(Ub(index)-Lb(index)).*rand;% Randomly
update a randomly selected (index) component  of the cooling object.
        end
    else
        newTO(i,:)=envTO(i-nTO/2,:)+(TO(i,:)-envTO(i-nTO/2,:))*exp(-
beta*t);
        if rand<p
            index=ceil(rand*nV);
            newTO(i,index)=Lb(index)+(Ub(index)-Lb(index)).*rand;% Randomly
update a randomly selected (index) component  of the cooling object.
        end
    end
end
```

The forth function is the *Replacement* function by which a fixed number (*STO_M*) of poor current objects is replaced with the memorized ones. Input arguments are the current set of thermal objects (*TO*), their corresponding objective function (*Fit*) and the penalized objective function (*PFit*) vectors, and the memorized objects (*TO_M*, *Fit_M*, and *PFit_M*). The output arguments are the *TO*, *Fit*, and *PFit* with replaced bad objects. The *Replacement* function is coded as follows:

```
% Replacing the poor objects with the memorized ones

function [TO,Fit,PFit]=Replacement(TO,Fit,PFit,TO_M,Fit_M,PFit_M)

nTO=size(TO,1);

helpTO=[TO;TO_M];helpFit=[Fit Fit_M];helpPFit=[PFit PFit_M];

[~,order]=sort(helpFit);

TO=helpTO(order(1:nTO),:);
Fit=helpFit(order(1:nTO));
PFit=helpPFit(order(1:nTO));
```

The fifth function is the *Memory* function by which the thermal objects memory matrix (*TO_M*) and its corresponding vectors of the objective function memory

(*Fit_M*) and the penalized objective function memory (*PFit_M*) are updated. The input arguments are *TO*, *Fit*, *PFit*, *TO_M*, *Fit_M*, and *PFit_M*, and the output arguments are *TO_M*, *Fit_M*, and *PFit_M*. The *Memory* function is coded in the follows. It should be noted that in the initialization phase, the memorizing should take place firstly. However, the *Memory* function will not be called for memorizing in the initialization phase because of its simplicity in this phase.

```
% Update the thermal objects memory matrix(TO_M) and its corresponding
vectors of objective function memory (Fit_M) and penalized objective
function memory (PFit_M).

function [TO_M,Fit_M,PFit_M]=Memory(TO,Fit,PFit,TO_M,Fit_M,PFit_M)

nTO=size(TO,1);
STO_M=size(TO_M,1);

for i=1:nTO
    for j=1:STO_M
        if PFit(i)<PFit_M(j)
            TO_M(j,:)=TO(i,:);
            Fit_M(j)=Fit(i);
            PFit_M(j)=PFit(i);
            break
        end
    end
end
```

TEO algorithm is coded in the following which is composed of three parts: initialization, algorithm cyclic body, and monitoring the results.

```
clc
clear

%% Initialization

% Define the properties of COP (tension/compression spring design problem).
nV=3; % Number of design variables.
Lb=[0.05 0.25 2]; % Lower bounds of design variables.
Ub=[2 1.3 15]; % Upper bounds of design variables.

% Define the parameters of the TEO algorithm.
nTO=20; % Number of Thermal Objects.
maxNFEs=20000; % Maximum Number of Objective Function Evaluations.
p=0.3; % With the probability of p a component of updated objects will be
regenerated randomly within the search space.
STO_M=5; % Size of the thermal objects memory.

% Randomly generate nTO number of thermal objects as the initial population
of the algorithm.
for i=1:nTO
    TO(i,:)=Lb+(Ub-Lb).*rand(1,nV); % Thermal objects matrix or matrix of
the initial candidate solutions or the initial population..
end

% Evaluate initial population (TO) calling the fobj function constructed in
the second chapter and form its corresponding vectors of objective function
(Fit) and penalized objective function (PFit). It should be noted that the
design vectors all are inside the search space.
for i=1:nTO
    [X,fit,pfit]=fobj(TO(i,:),Lb,Ub);
    TO(i,:)=X;
    Fit(i)=fit;
    PFit(i)=pfit;
end


% Form the thermal objects memory matrix(TO_M) and its corresponding
vectors of objective function memory (Fit_M) and penalized objective
function memory (PFit_M). It should be noted that in the initialization
phase it is not needed to utilize the Memory function and the memory simply
can be produced.
[value,index]=sort(PFit);
TO_M=TO(index(1:STO_M),:);Fit_M=Fit(index(1:STO_M));PFit_M=PFit(index(1:STO
_M));

%% Algorithm Body

NFEs=0; % Current Number of Objective Function Evaluations used by the
algorithm until yet.
NITs=0; % Number of algorithm iterations
maxNITs=maxNFEs/nTO; % Maximum Number of algorithm iterations

while NFEs<maxNFEs
    NITs=NITs+1; %Update the number of algorithm iterations.

    t=NITs/maxNITs; % Time is associated with the iteration number.

    % Determine the environmental temperature of the objects.
    envTO=Environment(TO,t);

    % Update the thermal objects by cooling or heating based on the
Newton's law of cooling.
```

```
    [newTO]=Newtonslaw (TO,envTO,Fit,PFit,p,t,Lb,Ub);

    % Evaluate the updated objects. It should be noted that in the TEO
algorithm the replacement strategy is not used except for a fixed number
(STO_M) of poor objects so that will be replaced with the memorized ones.
    for i=1:nTO
        [X,fit,pfit]=fobj(newTO(i,:),Lb,Ub);
        TO(i,:)=X;
        Fit(i)=fit;
        PFit(i)=pfit;
    end

    % Update the number of Objective Function Evaluations used by the
algorithm until yet.
    NFEs=NFEs+nTO;

    % Replace a fixed number (STO_M) of poor objects with the memorized
ones.
    [TO,Fit,PFit]=Replacement(TO,Fit,PFit,TO_M,Fit_M,PFit_M);


    % Update the thermal objects memory matrix(TO_M) and its corresponding
vectors of objective function memory (Fit_M) and penalized objective
function memory (PFit_M) utilizing the Memory function.
    [TO_M,Fit_M,PFit_M]=Memory(TO,Fit,PFit,TO_M,Fit_M,PFit_M);

    % Monitor the best memorized candidate solution (bestTO) and its
corresponding objective function (minFit) and penalized objective funtion
(minPFit).
    MinPFit=PFit_M(1);MinFit=Fit_M(1);bestTO=TO_M(1,:);

    % Display desired information of the iteration.
    disp(['NITs= ' num2str(NITs) '; MinFit = ' num2str(MinFit) '; MinPFit
= ' num2str(MinPFit)]);

    % Save the required results for post processing and visualization of
the algorithm performance.
    output1(NITs,:)=[MinFit,MinPFit,NFEs];
    output2(NITs,:)=[min(PFit),max(PFit),mean(PFit)];
    output3(NITs,:)=[bestTO,NFEs];
end

%% Monitoring the results
figure;
plot((1:1:NITs),output2(:,1),'g',(1:1:NITs),output2(:,2),'r--
',(1:1:NITs),output2(:,3),'b-.')
legend('min','max','mean');
xlabel('NITs');
ylabel('PFit');
```

15.4 Experimental Evaluation

TEO has four parameters which are as follows: maximum number of objective function evaluations (*maxNFEs*) as the stopping criteria of the algorithm, the number of thermal objects (*nTO*), the probability (*p*) for randomly regenerating one randomly selected component of the updated objects, and the size of the thermal objects memory (*STO_M*).

The simulation results show that:

Considering at least 15 numbers of thermal objects is required.

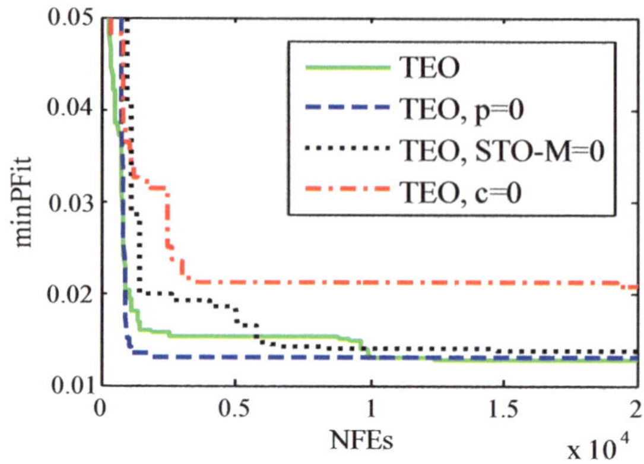

Fig. 15.3 Convergence histories of TEO with and without its memory and randomness mechanisms

TEO can work with a large number of thermal objects ($nTO > 50$), but one must consider significant size for the memory, at least one-quarter of it ($STO\_M > nTO/4$).

Ignoring the memory ($STO\_M = 0$), and/or ignoring the probability parameter ($p = 0$), and/or ignoring the incorporated randomness in determining the environmental temperature ($c = 0$) results in disrupting of the algorithm performance. Figure 15.3 monitors convergence histories of TEO ($nTO = 20$; $maxNFEs = 20{,}000$; $p = 0.3$; $STO\_M = 5$), and TEO ignoring the abovementioned items for single runs started from a fixed initial solution.

Considering p value minimally equal to 0.1 is required.

References

1. Kaveh A, Dadras A (2017) A novel meta-heuristic optimization algorithm: thermal exchange optimization. Adv Eng Softw 110:69–84
2. Kaveh A, Dadras A (2018) Structural damage identification using an enhanced thermal exchange optimization algorithm. Eng Optim 50:430–451
3. Kaveh A, Dadras A, Bakhshpoori T (2018) Improved thermal exchange optimization algorithm for optimal design of skeletal structures. Smart Struct Syst 21:263–278

Printed by Printforce, the Netherlands